MAKING
Pre-Algebra
COME
Alive

Student Activities & Teacher Notes

ALFRED S. POSAMENTIER

CORWIN PRESS, INC.
A Sage Publications Company
Thousand Oaks, California

Material from *Principles and Standards for School Mathematics,* copyright 2000 by the National Council of Teachers of Mathematics, is reprinted by permission of the NCTM.

For information:

Corwin Press, Inc.
A Sage Publications Company
2455 Teller Road
Thousand Oaks, California 91320
E-mail: order@corwinpress.com

Sage Publications Ltd.
6 Bonhill Street
London EC2A 4PU
United Kingdom

Sage Publications India Pvt. Ltd.
M-32 Market
Greater Kailash I
New Delhi 110 048 India

Library of Congress Cataloging-in-Publication Data

Posamentier, Alfred S.
  Making pre-algebra come alive: Student activities and teacher notes / by Alfred S. Posamentier.
     p. cm. — (Math assessment series)
  ISBN 0-7619-7594-2 (c) — ISBN 0-7619-7595-0 (p)
  1. Mathematics—Problems, exercises, etc. I. Title. II. Series.
  QA43 .P64 2000
  513—dc21   00-008376

00   01   02   03   04   05   10   9   8   7   6   5   4   3   2   1

Editorial Assistant:	Catherine Kantor
Production Editor:	Diana E. Axelsen
Editorial Assistant:	Victoria Cheng
Typesetter/Designer:	Technical Typesetting, Inc.
Designer:	Tracy Miller

CONTENTS

Introduction

Making Pre-Algebra Come Alive is a set of versatile enrichment exercises that cover a very broad range of mathematical topics and applications—from the Moebius strip to the googol. Several criteria have been used to develop the activities and to select the topics that are included. All of them bear heavily, and equally, on my concerns for curriculum goals and classroom management.

First and foremost, the activities are meant to be motivational. As much as possible, I want this book to achieve the goal of being attractive to people who thought they didn't like mathematics. To accomplish this, it is necessary for the activities to be quite different from what students encounter in their basal texts—different in both substance and form. This seems especially critical: No matter how excellently a basal text is being used, nearly every class experiences the "blahs." Unfortunately, this sort of boredom is often well entrenched long before the teacher and perhaps even the students are aware of it. Presenting activities on a regular basis gives the variety and change of pace needed to sustain interest in any subject.

With the number of topics you may have to cover during the normal school year, it may seem naïve or unrealistic to suggest introducing additional material. This brings me to the second criterion. Most of the activities in this volume can be used to enhance, reinforce, and extend the concepts and skills that already make up the better part of your curriculum and course goals. For some examples, see the activities in the problem-solving sections of this book. These clearly reinforce your work in several areas. Similarly, the arithmetic operations section provides enhancement as well as variety to your work with computation skills. Thus, the activities are designed as an aid to presenting the basic concepts of your course, as well as a set of motivational and enrichment activities. These objectives are completely consistent with the *Principles and Standards for School Mathematics* (NCTM, 2000).

Third, it was felt that each activity should have some use or merit beyond itself, a heuristic value. That is, the activities serve as door openers; they can be introductions to areas not usually treated in basal texts. The activities provide good practice in what you're trying to teach anyway, but they also greatly increase your students' awareness of the different directions to which these ideas can lead.

The Key: Problem Solving

Finally, the activities provide opportunities and incentives to hone problem-solving skills—not merely chapter-end exercises that are often called problems, but realistic problems such as your students will encounter in their everyday living and in later, nonmathematics school courses. Most of the activities begin by posing a problem that students find intriguing and that, at the outset, many students are unable to solve on their own. In working through the problem, however, the students discover they can tackle a much bigger monster than they thought they were capable of handling. Equally important, they find these problem-solving techniques are applicable to other areas.

The problem-solving orientation of these activities cannot be overemphasized. Those of you who are familiar with the NCTM Standards 2000 will find this book directly on target with most of the recommendations. Sometimes directly respondent, as with problem solving, and other times directly supportive, as with enrichment and applications, the *Math Alive Series* is a deliberate effort to meet the objectives of the Standards in a creative way.

Presenting the Activities

In pilot testing these activities, I worked with teachers who had very diverse mathematical preparation and who had to deal with a *wide* spectrum of class size, student ability, and class heterogeneity. Thus, it seemed very desirable to search for alternative means to present the activities. I discovered several. One or more of them should be useful in your situation.

The normal presentation, the one that best suits most classes, is to present the activity as a new lesson at the outset of a class period. In working through the student page, you'll find the accompanying Teacher's Notes explain the rationale for the entire activity, as well as provide anticipated student responses and questions. However familiar you feel with the mathematical topics presented, do not attempt to conduct a class session without first spending 20 or 30 minutes going over the Teacher's Notes. Both the student pages and the Teacher's Notes are highly compressed: A typical student page encompasses the concepts that four or five basal-text pages generally treat.

In some cases, the student activity can be handed out the day preceding class discussion. Your perusal of the activity will best determine when this is appropriate. In many other cases, you will find it best to discuss only the body of the student activity the day you pass it out, deferring the discussion of the Extension until the following day.

If your class is like many that I have encountered, you may wish to try peer teaching. This has many advantages for both you and your students if your classes have three to six really bright students. By giving both the student page and Teacher's Notes to one of these "stars," he or she can present the activity the following day to this group of above-average students. This allows you time to work with your average and below-average students to bring their skills up to par without boring the students who are already well on top of things. My experience has shown that students who are asked to present activities prepare very well. Their pride is at stake, and thus you can be sure they won't let you down.

The Extensions

The Extensions offer the greatest opportunity for flexibility in using the activities. Every activity in these volumes has one, but they differ. In some cases, they dip into more sophisticated mathematical concepts and should be considered as optional activities primarily for your better students. In other cases, the Extensions require no additional mathematical sophistication, but simply give an opportunity to explore the topic in greater detail. Your reading of the activity will quickly determine which is the case. Sometimes you may want to present the basic activity to the class and assign the Extension as homework for your better students. In all cases, you should think of the Extension as an element that allows you to tailor your mathematics program to best meet the needs and interests of all your students.

Selecting the Activities

This volume probably contains more activities than you'll be able to use in a single school year. The chapter introductions will assist you in selecting the activities best suited to your students' abilities and interests and offer some hints as to how they can be used. The activities have been divided into six categories. These categories are not at all arbitrary, but your study of them will show that considerable overlap is possible. For example, some of the binary activities could easily qualify as recreational, and some of the recreational fit well with your arithmetic operations teaching. Nevertheless, these descriptions or categorizations can provide an aid to organizing your use of the activities. It is possible in many classes to use all of the activities by selecting the easier ones for your slower students and using the more difficult ones as extra-credit work for your better students.

The following pages give an overview of the activities, category by category. A diamond (♦) follows the titles of the activities that are within the reach of your slower students. A star (⋆) indicates the activities that are probably best given only to your better students or given a more careful presentation to the general class. (Note that a ♦ doesn't mean that your better students won't like the activity; it simply means the activity is within the grasp of your mathematically less proficient ones.)

The NCTM Principles and Standards for School Mathematics – 2000

Each unit is tied in with one or more of the NCTM Standards presented in *Principles and Standards for School Mathematics* – 2000. As units are selected for use in the classroom, it is good to be aware of the Standards being employed. A simple numbering system is used to help make this identification simple and unobtrusive. At the start of each "Teacher Notes" section, the Standard number appropriate for that unit is indicated by a dot below the appropriate Standard number. These numbers correspond to the following list of standards:

1. Number and Operations Standard

Instructional programs from prekindergarten through grade 12 should enable all students to –

- understand numbers, ways of representing numbers, relationships among numbers, and number systems;
- understand meanings of operations and how they relate to one another;
- compute fluently and make reasonable estimates.

2. Algebra Standard

Instructional programs from prekindergarten through grade 12 should enable all students to –

- understand patterns, relations, and functions;
- represent and analyze mathematical situations and structures using algebraic symbols;
- use mathematical models to represent and understand quantitative relationships;
- analyze change in various contexts.

3. Geometry Standard

Instructional programs from prekindergarten through grade 12 should enable all students to –

- analyze characteristics and properties of two- and three-dimensional geometric shapes and develop mathematical arguments about geometric relationships;
- specify location and describe spatial relationships using coordinate geometry and other representational systems;
- apply transformations and use symmetry to analyze mathematical situations;
- use visualization, spatial reasoning, and geometric modeling to solve problems.

4. Measurement Standard

Instructional programs from prekindergarten through grade 12 should enable all students to –

- understand measurable attributes of objects and the units, systems, and processes of measurement;
- apply appropriate techniques, tools, and formulas to determine measurements.

5. Data Analysis and Probability Standard

Instructional programs from prekindergarten through grade 12 should enable all students to –

- formulate questions that can be addressed with data and collect, organize, and display relevant data to answer them;
- select and use appropriate statistical methods to analyze data;
- develop and evaluate inferences and predictions that are based on data;
- understand and apply basic concepts of probability.

6. Problem Solving Standard

Instructional programs from prekindergarten through grade 12 should enable all students to –

- build new mathematical knowledge through problem solving;
- solve problems that arise in mathematics and in other contexts;
- apply and adapt a wide variety of appropriate strategies to solve problems;
- monitor and reflect on the process of mathematical problem solving.

7. Reasoning and Proof Standard

Instructional programs from prekindergarten through grade 12 should enable all students to –

- recognize reasoning and proof as fundamental aspects of mathematics;
- make and investigate mathematical conjectures;
- develop and evaluate mathematical arguments and proofs;
- select and use various types of reasoning and methods of proof.

8. Communication Standard

Instructional programs from prekindergarten through grade 12 should enable all students to –

- organize and consolidate their mathematical thinking through communication;
- communicate their mathematical thinking coherently and clearly to peers, teachers, and others;
- analyze and evaluate the mathematical thinking and strategies of others;
- use the language of mathematics to express mathematical ideas precisely.

9. Connections Standard

Instructional programs from prekindergarten through grade 12 should enable all students to –

- recognize and use connections among different mathematical ideas;
- understand how mathematical ideas interconnect and build one another to produce a coherent whole;
- recognize and apply mathematics in contexts outside of mathematics.

10. Representation Standard

Instructional programs from prekindergarten through grade 12 should enable all students to –

- create and use representations to organize, record, and communicate mathematical ideas;
- select; apply, and translate among mathematical representations to solve problems;
- use representations to model and interpret physical, social, and mathematical phenomena.

ABOUT THE AUTHOR

Alfred S. Posamentier is Professor of Mathematics Education and Dean of the School of Education of the City College of the City University of New York. He is the author and coauthor of numerous mathematics books for teachers and secondary school students. As a guest lecturer, he favors topics regarding aspects of mathematics problem solving and the introduction of uncommon topics into the secondary school realm for the purpose of enriching the mathematics experience of those students. The development of this book reflects these penchants.

After completing his A.B. degree in mathematics at Hunter College of the City University of New York, he took a position as a teacher of mathematics at Theodore Roosevelt High School in the Bronx (New York), where he focused his attention on improving the students' problem-solving skills. He also developed the school's first mathematics teams (at both the junior and senior level) and established a special class whose primary focus was on mathematics problem solving and enrichment topics in mathematics.

For years, Dr. Posamentier has collected clever ways of introducing students to new concepts in mathematics. This collection of ideas prompted the development of this book. He is currently involved in working with mathematics teachers, locally and internationally, to help them better understand problem-solving strategies and alternative instructional strategies, so that they can comfortably incorporate them into their regular instructional program.

Immediately upon joining the faculty of the City College (after having received his masters' degree there), he began to develop inservice courses for secondary school mathematics teachers, including such special areas as recreational mathematics, problem solving in mathematics, and instructional alternatives for the classroom.

Dr. Posamentier received his Ph.D. from Fordham University (New York) in mathematics education and has since extended his reputation in mathematics education to Europe. He is an Honorary Fellow at the South Bank University (London, England). He has been visiting professor at several Austrian, British, and German universities, most recently at the University of Vienna and at the Technical University of Vienna. At the former, he was a Fulbright Professor in 1990.

In recognition of his outstanding teaching, the City College Alumni Association named him Educator of the Year and had a day (May 1, 1993) named in his honor by the City Council President of New York City. In 1994, he was awarded the National Medal of Honor from the Austrian government, and in 1999, by an act of the Austrian Parliament he was awarded an Austrian University Professor title by the President of the Republic of Austria.

Naturally, with his penchant for motivating students towards mathematics, he has been very concerned that students have a proper introduction to mathematics from an entertaining point of view. This interest motivated the development of this book.

CHAPTER 1

Number Theory and Arithmetic Operations

- The Fascinating Number 9
- Symmetric Multiplication
- Multiplication—Still More Ways
- Divisibiliti
- Triangular Numbers ★
- Prime After Prime ★

In most secondary school curricula, pre-algebra is the last course in which specific attention is paid to multiplication, division, squares, cubes, and primes. Thereafter, these skills and operations are pretty much taken for granted. Thus, many teachers will regard this section as the heart of the pre-algebra activities volume, and in many respects, it is, though for somewhat broader reasons than some might suspect.

The six activities in this section provide excellent drill in the above-mentioned skills. On another level, they generate considerable motivation—both from the aesthetic appeal that the patterns of numbers form and from the "gee whiz" surprises that pop up. The fact that they're fun is more than incidental. Of greatest importance, however, is that these activities explore the operational basis of our number system. These are the activities that will intrigue your potential math majors. However, "The Fascinating Number 9" and "Divisibiliti" usually appeal to everyone.

"The Fascinating Number 9" is invariably a pleasant surprise to students because at first glance it appears that they're in for a lot of tedious calculation. Not only will students find the calculations on this sheet go a lot faster than they expect, they'll learn some useful shortcuts that are more generally applicable. The Extension leads students to discover how the quirks of 9 crop up in other bases as well.

"Symmetric Multiplication" and "Multiplication—Still More Ways" probe into what's actually going on during simple multiplication exercises. Again, students will be fascinated by the visual arrays of numbers that look "too perfect" to be correct. The Extension to "Multiplication—Still More Ways" gives the secret to an old parlor trick that many students will enjoy.

In terms of practical applications, "Divisibiliti" is probably the most useful of the activities in this group. The usefulness of this activity's information and the ease with which it can be acquired assure plenty of motivation. However, only your very best students will be able to deduce the rules asked for in the Extension. Therefore, as suggested in the Teacher's Notes, you'll want to carefully walk your students through all of the examples given on the student page.

There are many people who are not receptive or sensitive to the aesthetic side of mathematics—just as one person may have no ear for music or another no eye for painting. No amount of instructional effort has any effect here. However, for those who do appreciate the beauty and the quirks of math, "Triangular Numbers" and "Prime After Prime" will be a treat.

Primes are treated in nearly all pre-algebra texts. You may find, however, that the treatment in "Prime After Prime" is more systematic and clearer. Both "Prime After Prime" and "Triangular Numbers" are surely among the most open-ended of these activities and can easily occupy several day's work if you and your students are so inclined.

The Fascinating Number 9

Quickly scan the problems listed below. It may look as though we've given you enough work to keep you busy multiplying and adding all day. But you can probably do all of these problems in less than 15 minutes—because we are working with the fascinating number nine. Fill in the blanks below and you'll see why! Use a separate sheet of paper for whatever calculations you need to make.

1. 12345679 × 9 = _____

    12345679 × 18 = _____

    12345679 × 27 = _____

    12345679 × 36 = _____

    12345679 × 45 = _____

    12345679 × 54 = _____

    12345679 × 63 = _____

    12345679 × 72 = _____

    12345679 × 81 = _____

2. 987654321 × 9 = _____

    987654321 × 18 = _____

    987654321 × 27 = _____

    987654321 × 36 = _____

    987654321 × 45 = _____

    987654321 × 54 = _____

    987654321 × 63 = _____

    987654321 × 72 = _____

    987654321 × 81 = _____

3.
$$1 \times 9 + 2 = \rule{3cm}{0.4pt}$$
$$12 \times 9 + 3 = \rule{3cm}{0.4pt}$$
$$123 \times 9 + 4 = \rule{3cm}{0.4pt}$$
$$1234 \times 9 + 5 = \rule{3cm}{0.4pt}$$
$$12345 \times 9 + 6 = \rule{3cm}{0.4pt}$$
$$123456 \times 9 + 7 = \rule{3cm}{0.4pt}$$
$$1234567 \times 9 + 8 = \rule{3cm}{0.4pt}$$
$$12345678 \times 9 + 9 = \rule{3cm}{0.4pt}$$

4.
$$9 \times 9 + 7 = \rule{3cm}{0.4pt}$$
$$98 \times 9 + 6 = \rule{3cm}{0.4pt}$$
$$987 \times 9 + 5 = \rule{3cm}{0.4pt}$$
$$9876 \times 9 + 4 = \rule{3cm}{0.4pt}$$
$$98765 \times 9 + 3 = \rule{3cm}{0.4pt}$$
$$987654 \times 9 + 2 = \rule{3cm}{0.4pt}$$
$$9876543 \times 9 + 1 = \rule{3cm}{0.4pt}$$
$$98765432 \times 9 + 0 = \rule{3cm}{0.4pt}$$

5.
$$999999 \times 2 = \rule{3cm}{0.4pt}$$
$$999999 \times 3 = \rule{3cm}{0.4pt}$$
$$999999 \times 4 = \rule{3cm}{0.4pt}$$
$$999999 \times 5 = \rule{3cm}{0.4pt}$$
$$999999 \times 6 = \rule{3cm}{0.4pt}$$
$$999999 \times 7 = \rule{3cm}{0.4pt}$$
$$999999 \times 8 = \rule{3cm}{0.4pt}$$
$$999999 \times 9 = \rule{3cm}{0.4pt}$$

6.
$$9 \times 9 = \rule{3cm}{0.4pt}$$
$$99 \times 99 = \rule{3cm}{0.4pt}$$
$$999 \times 999 = \rule{3cm}{0.4pt}$$
$$9999 \times 9999 = \rule{3cm}{0.4pt}$$
$$99999 \times 99999 = \rule{3cm}{0.4pt}$$
$$999999 \times 999999 = \rule{3cm}{0.4pt}$$
$$9999999 \times 9999999 = \rule{3cm}{0.4pt}$$

You won't often run into combinations such as the numbers shown above. But the number 9 sometimes offers a good shortcuts for fast multiplication. Multiply 53×99 as you normally would. See how much faster it is to recall that $99 = 100 - 1$. Thus, $53 \times 99 = 53(100 - 1) = 5300 - 53 = 5247$. Now try 47×999 and 67×990.

EXTENSION! In Problem 1 above the number 8 is omitted. Why? If we multiplied 7 times a base 8 sequences, what number would be left out of the sequence? A base 6 sequence?

Teacher's Notes for The Fascinating Number 9

This activity offers a recreational, entertaining examination of some of the properties of the number nine. Most students are delighted by the patterns formed by these multiplications and are thus motivated to investigate number properties and the base-ten system.

				NCTM Standards					
1	2	3	4	5	6	7	8	9	10
•	•				•	•	•		

Presenting the Activity

Before students work through the problems, discuss some of the properties of nine in multiplication and division. Write the products of nine and the integers 1 through 9 $(9, 18, \ldots, 72, 81)$ on the chalkboard. The students should easily see that the sum of the digits of each product is 9. Thus, a number is divisible by 9 if the sum of its digits is divisible by 9. Also note that the one's digits decrease from 9 to 1 while the ten's digits increase from 0 to 8. Point out that because $9 = 10 - 1$, the properties of nine are closely related to the properties of ten. For example, $9 \cdot 2 = (10 - 1) \cdot 2 = 20 - 2 = 18$.

Now have students work the problems. Encourage them to look for a pattern after they have multiplied two or three problems in a series. Also ask students to investigate *why* these patterns occur.

In Problem 1, note that

$$12345679 \times 9 = 12345679(10 - 1).$$

This is the subtraction problem

$$
\begin{array}{r}
123456790 \\
-12345679 \\
\hline
111111111
\end{array}
$$

In the rest of the problems in the series, the first problem is multiplied by $2, 3, 4, \ldots, 9$. Thus, the answer to the first problem $(111,111,111)$ is multiplied by $2, 3, 4, \ldots, 9$.

Problem 2 is similar to Problem 1 in that

$$
\begin{aligned}
987654321 \times 9 &= 987654321(10 - 1) \\
&= 9876543210 - 987654321 \\
&= 888888889.
\end{aligned}
$$

This answer is then multiplied by $2, 3, 4, \ldots, 9$. The duplication of the 8s causes the duplication of digits throughout the series. The descending digits at the ends of the numbers and the ascending digits at the beginnings follow the pattern of multiplying 9 by the integers 1 through 9.

In Problem 3, use the last problem in the series to show why the pattern occurs:

$$12345678 \times 9 + 9 = 12345678(10 - 1) + 9$$
$$= 123456780 - 12345678 + 9$$
$$= 123456789 - 12345678$$
$$= 111111111.$$

Multiplying by 10 adds a zero to the consecutive increasing digits. Then, addition of the next consecutive digit gives a number with one more digit than the original number.

In Problem 4, use one of the middle problems in the series to show the pattern:

$$98765 \times 9 + 3 = 98765(10 - 1) + 3$$
$$= 987650 - 98765 + 3$$
$$= 987653 - 98765$$
$$= 888888.$$

Multiplication by 10 adds a zero to the consecutive decreasing digits. Because the answer is to be all 8s, a number must be added so that the last digit of the answer will be an 8 when the last digit of the original number is subtracted from 10.

In Problem 5,

$$999999 \times n = (1000000 - 1)n = 1000000n \quad n.$$

The initial digits and final digits follow the pattern of the products of 9 and the integers 1 through 9, i.e., 18, 27, etc. Note that this pattern would be the same regardless of the number of 9s in the factor. The number of 9s in the answer is one less than the number in the factor.

For Problem 6,

$$9 \times 9 = (10 - 1)^2 = 100 - 20 + 1 = 101 - 20 = 81,$$
$$99 \times 99 = (100 - 1)^2 = 10,000 - 200 + 1$$
$$= 10,001 - 200 = 9801,$$
$$999 \times 999 = (1000 - 1)^2 = 1,000,000 - 2000 + 1$$
$$= 1,000,001 - 2000 = 998,001.$$

Encourage students to find other patterns using nine and to find other patterns in the six problems above. (For example, in Problem 1, there are nine 2s in the second answer and one factor is 9×2).

For 47×999 and 67×990, the students' work should be similar to the following:

$$47 \times 999 = 47(1000 - 1)$$
$$= 47,000 - 47$$
$$= 46,953;$$
$$67 \times 990 = 67(1000 - 10)$$
$$= 67,000 - 670$$
$$= 66,330.$$

Extension

Have students multiply in Problem 1 with the 8 included in the series:

$$123456789 \times 9 = 123456789(10 - 1)$$
$$= 1234567890 - 123456789$$
$$= 1111111101.$$

In the subtraction $1234567890 - 123456789$, the difference between the last two digits $(90 - 89)$ is only 1. For the pattern to work, the difference must be 11 (i.e., $90 - 79$). Thus, there is "a base too many." In base 10 there is 10 more (89 instead of 79), than allows the pattern to work. By omitting the number in the consecutive digits that represents the base minus 2, the pattern works. Thus in base 8 sequence, the 6 would be omitted. In a base 6 sequence, the 4 would be omitted.

Symmetric Multiplication

If you've worked through "The Fascinating Number 9," you've seen that some pretty funny looking patterns can be produced by multiplication. Here you'll see some that are even more weird. Start by multiplying

$66666 \times 66666 =$ _____ , $2222 \times 2222 =$ _____ , $333 \times 888 =$ _____ .

Now let's look at some different ways to multiply and see what we can learn. Read through the following multiplications and see if you can figure why they turn out as they do.

```
        66666              2222              888
      X 66666            X 2222            X 333
         36                 04               24
        3636               0404             2424
       363636             040404           242424
      36363636           04040404           2424
     3636363636          040404              24
      36363636            0404             295704
       363636              04
        3636             4937284
         36
   4444355556
```

Now this surely isn't the way you were taught multiplication a few years ago. It's time to scratch your heads—being careful not to get splinters in your fingers. Why do these multiplications work this way?_____

_____

_____

_____

If you've come this far, you can probably explain examples such as

```
      8888                66666
    X 3333              X 66666
       8                   6
      888                 666
     88888               66666
    8888888             6666666
    9874568            666666666
        X 3            740725926
   29623704                 X 6
                      4444355556
```

What's going on here? That is, how does this work?_____

_____

_____

EXTENSION! Will this form of multiplication work if the numbers don't contain the same number of digits? Try 4444 × 666 and 22222 × 777.

Teacher's Notes for Symmetric Multiplication

Many teachers like this activity because of its versatility in application. Even your slower students can appreciate and find amusement from the visual arrays that these problems yield. Your faster students, of course, are provided with an opportunity to examine decimal multiplication operations in depth. Ask students to show their work.

——————————————— NCTM Standards ———————————————

1	2	3	4	5	6	7	8	9	10
•	•								

Presenting the Activity

It's usually a good idea to have students perform the first three indicated multiplications *before* passing out the student activity sheets. The students' calculations will look something like

```
      66666          2222           888
    X 66666        X 2222         X 333
     399996          4444          2664
     399996          4444          2664
     399996          4444          2664
     399996          4444        295704
     399996        4937284
  4444355556
```

Then have the students record their answers on the student page and compare their work to the symmetric multiplications shown. A lot of space is provided for their answers to why the presented patterns can work this way, and most of your discussion should be on this point. What students will see, for example, is that $6 \times 6 = 36$. Carrying 3 in subsequent multiplications gives 39, which in turn produces a lot of $9 + 6$s. Have students change the pattern of 36s on the student page to the usual multiplication shown above. They can do this by adding some of the 6s and 3s to get 9s. Thus, the sum of each column of digits is the same; the digits within each column have simply been rearranged.

This multiplication is sometimes called *rhombic*, because of the shape of the pattern. Similarly, the examples at the bottom of the student page are called *triangular* multiplication. The use of these terms is left to your discretion.

Now consider the two examples at the bottom of the page. Students should see that the triangular array is produced by factoring 3 out of 3333 to produce 8888×1111. The array you would normally produce is

```
      8888
    X 1111
      8888
     8888|
    8888||
   8888↓↓↓
```

11

If you simply "push down" the three columns on the right (as the arrows indicate), you get the pattern that is shown on the student page. Students can easily see that the subsequent addition is not affected. The product of 8888 and 1111 is then multiplied by 3. If necessary, show another example so students can see how the triangular pattern can be changed to the usual multiplication pattern by rearranging the last three columns.

Extension

This presents both a visually and intellectually interesting phenomenon. The multipliers that have two or fewer digits less than the multiplicand are compensated for in the following way: Count down a number of rows equal to the number of digits in the multiplier. Then draw diagonals as shown here. "Erase" everything below the diagonals and the products are miraculously achieved.

```
      44444                    22222
    X 666                    X 777
      24                       14
      2424                     1414
      242424                   141414
      24242424                 14141414
      2424242424               1414141414
      242424242424             141414141414
      24242424                 14141414
      2424                     1414
      24                       14
    29599704                 17266494
```

Multiplication—Still More Ways

You've been multiplying numbers for several years now, but have you ever stopped to think about how the system works? Multiply 43×92 as you normally do.

$$\begin{array}{r} 92 \\ \times\ 43 \\ \hline \end{array}$$

To obtain your answer, what you actually do is multiply 92 by $(40 + 3)$. This gives you $3680 + 276 = 3956$.

Now let's consider another method, called the *doubling method*, and see how (and why) it works. Make two columns, one headed by "1" and the other headed by "92."

★	1	92
★	2	184
	4	368
★	8	736
	16	1472
★	32	2944

Fill in the table, doubling each time. Why do we stop at 32?_____

_____

Now add the numbers in the starred rows. What is the sum of the starred numbers in the *first* column?_____ What is the sum of the starred numbers in the *second* column?_____ Why have some of the rows been starred and not others?_____

_____

Still another method is called *lattice multiplication*. Again using our example of 43×92, we build an array such as:

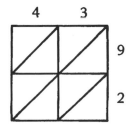

Begin by multiplying 3×9 in the upper-right corner to get ⬚.

13

Continue with 4×9, 4×2, and 3×2. Your lattice then looks like

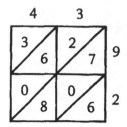

Notice that when your product is a single digit (as with 4×2 and 3×2), a zero is put into the space where you would normally put the ten's digit.

Now add the four diagonals as indicated by the arrows. (Start at the lower right, moving up and to the left.)

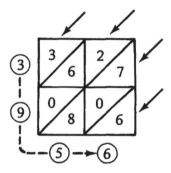

Your answer, 3956, appears in the circled numbers. Try some the following additional examples.

36 X 75   **56 X 14**   **23 X 67**

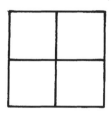

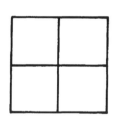

EXTENSION! Multiply the numbers

$19 \times 21 = $ _____ , $32 \times 28 = $ _____ ,

$43 \times 37 = $ _____ , $56 \times 64 = $ _____ .

Can you spot a pattern here? If you can, you can make people think you're a real whiz at rapid multiplication!

Teacher's Notes for Multiplication—Still More Ways

This activity, as is the case with the other units in this category, is not intended as drill or practice in multiplication. These activities offer an exploration of how the decimal system works. The Extension provides an often handy multiplication shortcut that is also an impressive parlor trick. Ask students to show their work.

―――――――――――――――――― NCTM Standards ――――――――――――――――――

1	2	3	4	5	6	7	8	9	10
•	•				•			•	

Presenting the Activity

None of your students will have any difficulty multiplying 43×92 and most will recognize that moving the 368, (4×92) one place to the left has the effect of making it 40×92.

In the doubling method, students stop doubling at 2^5 because the next number, 64, is greater than 43, the multiplicand. Adding the starred numbers in the two columns gives 43 and 3956. The starred powers of 2, of course, yield binary 43, 101011. You may wish to give a few more examples to assure that students understand why this works.

The lattice method will be familiar to students acquainted with Napier's rods and the principle involved here is the same. Be sure that your students notice that the "1" in the sum of the second diagonal is carried over to the third diagonal. Thus, adding the elements of the third diagonal yields 9 instead of 8. Completed lattices for the other student problems are

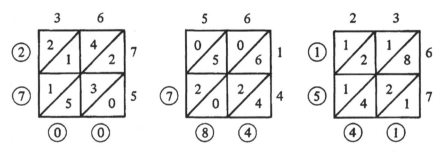

Students may wonder if this method works for numbers with more than two digits and you can assure them that it does:

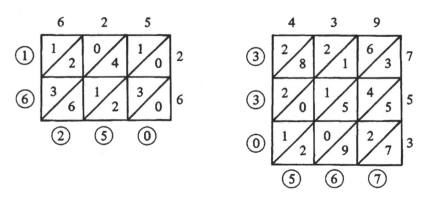

15

Extension

Many students will not spot the pattern on their own. When it's pointed out, however, they should have little trouble seeing that the Extension is based on the fact that

$$(a + b)(a - b) = a^2 - b^2.$$

Thus,

$$19 \times 21 = (20 - 1)(20 + 1) = 400 - 1 = 399,$$
$$32 \times 28 = (30 + 2)(30 - 2) = 900 - 4 = 896,$$
$$43 \times 37 = (40 + 3)(40 - 3) = 1600 - 9 = 1591,$$
$$56 \times 64 = (60 - 4)(60 + 4) = 3600 - 16 = 3584.$$

The "trick," then, to making yourself look like a calculating genius is to select numbers slightly (and equally) above and below easy squares.

Divisibiliti

Look at the numbers in the following list. Without actually dividing, can you tell which ones are divisible by 2, 3, or 5?

    a) 792 is divisible by_____ but not by_____.

    b) 835 is divisible by_____ but not by_____.

    c) 356 is divisible by_____ but not by_____.

    d) 3890 is divisible by_____ but not by_____.

    e) 693 is divisible by_____ but not by_____.

    f) 743 is divisible by_____ but not by_____.

The 2s and 5s are pretty easy; any even number is divisible by 2 and any number that ends in a 5 or 0 is divisible by 5. The 3s are a bit tougher. Look at the following numbers:

  924   651   72   105   1014   339   270   87.

They're all divisible by 3. See if you can discover a rule for divisibility by 3. A number is divisible by 3 if:_____

_____

Let's consider divisibility by the other numbers, starting with 4. Perhaps the easiest method is simply to divide by 2 twice. That is, after dividing by 2 if you still have an even number, then the original number is divisible by 4. Look at the numbers

  936   572   200   556   3328   740   5100   8344

that are all divisible by 4 and state a general rule for divisibility by 4:_____

_____

This rule can be extended to 8 by_____

_____

Since you've already come up with a rule for divisibility by 3, it's an easy step to find a rule for 6. A number is divisible by 6 if it's divisible by 3 and_____.
Now let's look at divisibility by 9. The numbers

405   36   522   324   873   747   9756   3375,

which are each divisible by 9, should allow you to state a rule for divisibility by 9:

_____

EXTENSION! Study the following sets of numbers. Set A contains numbers that are divisible by 7. Set B contains numbers that are divisible by 11. What are the rules for divisibility by 7 and 11?

Set A				Set B			
259	595	371	798	198	407	649	1012
336	1092	1253	504	286	1694	847	3531

Teacher's Notes for Divisibiliti

This unit provides a very useful spot check of divisibility. The activity contains problems that at first glance appear difficult. The beauty of it is the ease with which most of the problems can be solved. Ask to students to show their work.

				NCTM Standards					
1	2	3	4	5	6	7	8	9	10
•					•			•	

Presenting the Activity

The title is not a printer's error; it's simply a bit of Italian-flavored whimsy that yields a 12-letter word using for all the even-numbered letters. The caption on the figure indicates a man saying "Won't go" (literally "incorrect") to $143 \div 7$.

Without too much head-scratching, your students will see that 792 is divisible by 2 and 3, but not by 5; 835 by 5, but not by 2 or 3; 356 by 2, but not by 3 or 5; 3890 by 2 and 5, but not by 3; 693 by 3, but not by 2 or 5; and 743 by none of them. The rules for divisibility by 2 and 5 are given on the student page and shouldn't require much discussion.

There are eight examples given for divisibility by 3. These should be enough to show that a number is divisible by 3 if the sum of its digits is divisible by 3.

The step to divisibility by 4 is simple. As mentioned on the student page, the simplest test is just dividing by 2 twice. Put another way, if the last two digits form a number divisible by 4, the entire number is divisible by 4. (*Note*: Because of the test for divisibility by 3, a few students may make the mistake of thinking we're talking about the *sum* of the last two digits. Obviously, we're not.) Some students may find another way to express the rule: If the last digit is a 0, 4, or 8, the entire number is divisible by 4 if the next-to-last digit is even; if the last digit is a 2 or a 6, the entire number is divisible by 4 if the next-to-last digit is odd.

A number is divisible by 8 if it's divisible by 4 and again by 2. Analogous to the rule for divisibility by 4, a number is divisible by 8 if its last three digits (considered as a number) is divisible by 8. Ask your students why three digits are needed for this test, rather than just two. (All multiples of 100 are divisible by 4, but only even multiples of 100 are divisible by 8.) Then see if they can form a rule analogous to the last rule given for divisibility by 4.

A number is divisible by 6 if it is even and also divisible by 3. In the next problem, students will discover that, similar to the rule for 3, a number is divisible by 9 if the sum of its digits is divisible by 9. An easy explanation for why this is so is given in the "Digit Problems Revisited" activity in *Making Algebra Come Alive.*

Extension

To test for divisibility by 7, delete the last digit of the number, double it, and then subtract it from the number that remains. If the number that is obtained is 0 or divisible by 7, the original number is divisible by 7. Note that if the resulting number is still unwieldy, the test can be easily repeated. For example:

$$
\begin{array}{r}
59\text{⑤} \\
-\ 10 \\
\hline
49
\end{array}
\qquad
\begin{array}{r}
37\text{①} \\
-\ 2 \\
\hline
35
\end{array}
\qquad
\begin{array}{r}
125\text{③} \\
-\ 6 \\
\hline
11\text{⑨} \\
-\ 18 \\
\hline
-\ 7
\end{array}
\qquad
\begin{array}{r}
109\text{②} \\
-\ 4 \\
\hline
10\text{⑤} \\
-\ 10 \\
\hline
0
\end{array}
$$

The reason this rule works is that you are actually subtracting multiples of 7. As you can see in the preceding examples, subtracting 10 from 59 is the equivalent of subtracting 105 from 595 and dividing by 10. Subtracting 2 from 37 is the equivalent of subtracting 21 from 371 and dividing by 10. Because we've been subtracting integral multiples of 7, if the remainder is evenly divisible by 7, the original number must be divisible by 7. Demonstrate several examples for your students.

The same procedure can be used for other primes. The only difference is that the last digit is multiplied by different numbers before subtracting (to assure that you're subtracting integral multiples of the divisor). A table of some of these follows below. In a few cases the multipliers are large enough to nullify the value of the test.

To test divisibility by	7	11	13	17	19	23	29	31	37	41
Multiply the last digit by	2	1	9	5	17	16	26	3	11	4

An alternate test for divisibility by 11 is much easier. Consider the sums of the even and odd digits of the number. Compare the two sums. If their difference is either 0 or divisible by 11, the original number is divisible by 11. For example,

198:	$1 + 8 = 9$; $9 - 9 = 0$;
407:	$4 + 7 = 11$; $11 - 0 = 11$;
847:	$8 + 7 = 15$; $15 - 4 = 11$;
3531:	$3 + 3 = 6$, $5 + 1 = 6$; $6 - 6 = 0$.

Alternate Extension

The principles and operations of the student page Extension are easily grasped by most students *when examples are demonstrated on the chalkboard*. However, very few students are able to discover these on their own. Thus, assigning the Extension as independent study usually just amounts to handing out frustration. A better course is to discuss the extension in class, then ask the students to devise tests for divisibility by several composite numbers. Because they've already seen examples in testing divisibility by 4, 6, and 8, they can generate a table such as

To be divisible by	6	10	12	15	18	21	24	26	28
The number must be divisible by	2, 3	2, 5	3, 4	3, 5	2, 9	3, 7	3, 8	2, 13	4, 7

Triangular Numbers

In mathematics, we sometimes see patterns determined by numerical relationships. Sometimes the patterns result from geometric relationships. With triangular numbers, both of these factors produce a pattern. Let's start by completing the equations

$$
\begin{aligned}
1^3 &= \quad 1 \quad = 1^2, \\
1^3 + 2^3 &= \quad 9 \quad = 3^2, \\
1^3 + 2^3 + 3^3 &= \underline{\hspace{1.5cm}} = (\underline{\hspace{0.7cm}})^2, \\
1^3 + 2^3 + 3^3 + 4^3 &= \underline{\hspace{1.5cm}} = (\underline{\hspace{0.7cm}})^2, \\
1^3 + 2^3 + 3^3 + 4^3 + 5^3 &= \underline{\hspace{1.5cm}} = (\underline{\hspace{0.7cm}})^2.
\end{aligned}
$$

Now look at the numbers in the last column. Each number can be drawn as a triangular arrangement of dots. The first three numbers are

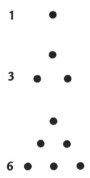

Draw the other two:

21

We call these numbers *triangular* numbers. What are the next three triangular numbers?_____

Triangular numbers have some unusual properties. For example, look at the pattern

$$1 + 3 = 2^2$$
$$3 + 6 = 3^2$$
$$6 + 10 = 4^2$$
$$10 + 15 = 5^2$$

What are the next three steps in the pattern?

_____

What "rule" describes the pattern?_____

_____

Notice the differences between consecutive triangular numbers. Continue the pattern for the next two triangular numbers.

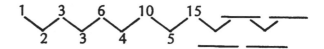

Triangular numbers: 1   3   6   10   15 ___ ___

Differences: 2   3   4   5 ___ ___

Triangular numbers can also help us get some odd squares. Complete the following equations:

$$8 \times 1 + 1 = 3^2, \qquad 8 \times 10 + 1 = \underline{\hspace{1.5cm}},$$
$$8 \times 3 + 1 = 5^2, \qquad 8 \times 15 + 1 = \underline{\hspace{1.5cm}},$$
$$8 \times 6 + 1 = 7^2, \qquad \underline{\hspace{1.5cm}} = \underline{\hspace{1.5cm}},$$
$$\underline{\hspace{1.5cm}} = \underline{\hspace{1.5cm}}.$$

EXTENSION! Every positive integer can be written as the sum of three (or fewer) triangular numbers. Express 17, 23, 37, and 46 as the sum of three (or fewer) triangular numbers.

Teacher's Notes for Triangular Numbers

Patterns of numbers occur frequently in mathematics as shown by other activities in this volume ("The Fascinating Number 9," "Symmetric Multiplication" and "Scamps"). Triangular numbers are particularly interesting because of the variety of both numerical and geometric patterns. Students will find that each pattern leads to a new relationship or discovery about numbers. Ask student to show their work.

NCTM Standards

1	2	3	4	5	6	7	8	9	10
•	•	•			•				

Presenting the Activity

Begin by briefly reviewing squares, cubes, and square roots. Then have students complete the first pattern:

$$1^3 = 1 = 1^2,$$
$$1^3 + 2^3 = 9 = 3^2,$$
$$1^3 + 2^3 + 3^3 = 36 = 6^2,$$
$$1^3 + 2^3 + 3^3 + 4^3 = 100 = 10^2,$$
$$1^3 + 2^3 + 3^3 + 4^3 + 5^3 = 225 = 15^2.$$

At first glance, the pattern appears to be only numerical. The next paragraph shows it to be geometric, too. The dot arrangements for 10 and 15 are

The next three triangular numbers after 15 are 21, 28, and 36. Ask students how they can find these numbers. Some students may continue the numerical pattern of sums of consecutive cubes. Others may use the geometric pattern and simply add a row of dots to the picture of the preceding triangular number. Others may notice that the differences between consecutive triangular numbers increase by 1. This pattern of differences is discussed a bit farther down the student page.

There are many unusual properties of triangular numbers. The next pattern shows that the sum of two consecutive triangular numbers produces a square number. If the triangular-number sums are taken in order, the square numbers occur in order. Thus, the next three steps in the pattern are

$$15 + 21 = 6^2,$$
$$21 + 28 = 7^2,$$
$$28 + 36 = 8^2.$$

Dot arrangements can be used to show why this pattern works:

$$6 \quad + \quad 10 \quad = \quad 16 \text{ or } 4^2$$

Pythagoras considered both square and triangular numbers (he called the latter *holy tetractys*), and believed they had special properties. You can show how any square number represented by dots can be divided by a diagonal line into two consecutive triangular numbers:

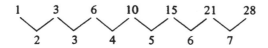

The next property of triangular numbers presented on the student page is the previously mentioned pattern of differences. The completed pattern is

1		3		6		10		15		21		28
	2		3		4		5		6		7	

If you put the pattern on the chalkboard, students can easily see adding down and to the right gives the next number in the top row, and the sum of any two adjacent numbers in the top row gives the square of the number below and between them.

This pattern gives rise to another numerical pattern for finding the triangular numbers:

$$1 = 1$$
$$1 + 2 = 3$$
$$1 + 2 + 3 = 6$$
$$1 + 2 + 3 + 4 = 10$$
$$1 + 2 + 3 + 4 + 5 = 15.$$

Notice that the combination of this pattern and the pattern of cubes at the top of the student page means

$$1^3 + 2^3 + 3^3 + 4^3 + 5^3 = (1 + 2 + 3 + 4 + 5)^2.$$

The completed pattern of odd squares is

$$8 \times 1 + 1 = 3^2$$
$$8 \times 3 + 1 = 5^2$$
$$8 \times 6 + 1 = 7^2$$
$$8 \times 10 + 1 = 9^2$$
$$8 \times 15 + 1 = 11^2$$
$$8 \times 21 + 1 = 13^2$$
$$8 \times 28 + 1 = 15^2.$$

Extension

Often more than one combination of triangular numbers can be used to express a given integer. For example, both $15+1+1$ and $10+6+1$ are equal to 17. For others, such as $23 = 21+1+1$, only one combination is possible. For 37, $36+1$ is one possibility and $28+6+3$ is another. For 46, $45+1$ is a possibility, $36+10$ is another, and $28+15+3$ is another.

Every positive integer can also be written as the sum of square numbers. In this case, four (or fewer) square numbers are necessary. If time permits, have students experiment with a few of these. For example, $23 = 9+9+4+1$, $37 = 36+1$, and $46 = 36+9+1$.

Prime After Prime

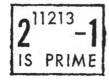 

COURTESY OF PAUL T. BATEMAN, DEPARTMENT OF MATHEMATICS, UNIVERSITY OF ILLINOIS.

At one time, the mathematics department at the University of Illinois used the postmark above on all the department's mail. Until 1971, $2^{11213} - 1$ was the largest prime number known. (Recall that a prime number has no factors except 1 and itself). It would be quite difficult without using a computer to prove that $2^{11213} - 1$ is a prime number—it contains 3376 digits.

Prime numbers have been of interest to mathematicians for thousands of years. A method for finding all the prime numbers less than 100 was discovered more than 2000 years ago. We can also use this method to find the factors of the composite numbers less than 100.

Use a chart with the whole numbers from 2 to 100. Put a "P" for prime in the box with the 2. Then put a "2" in each box that contains a multiple of 2. Put a "P" in the box with the 3. Then put a "3" in each box that contains a multiple of 3. The box with the 4 has the factor 2 in it, so you know 4 is not prime. Put a "4" in each box that contains a multiple of 4. Continue in this way until the chart is complete. The first row is

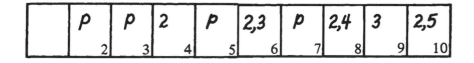

In addition to showing all the primes, your chart shows all the factors of the composite numbers. Thus, all the factors of 12 are 1, 2, 3, 4, 6, and 12. (Remember, 1 and the number itself are factors.)

What are all the factors of 54?_____

What are all the factors of 90?_____

You can also use the chart to make a "factor tree." This shows the prime factorization of a number. Let's try 24. First find any two factors whose product is 24, say 2 and 12. We know 2 is prime, but 12 is not. So find the box with 12 in it and find two factors whose product is 12, say 2 and 6. We know 2 is prime and 6 is not, so we find the factors of 6. The only factors of 6 are 2 and 3, both primes, so the prime factorization of 24 is $2 \times 2 \times 2 \times 3$. The completed factor

tree is

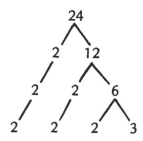

What is the prime factorization of 72?_____

How would you use the chart to find all the factors of a number greater than 100?

_____

What are all the factors of 156?_____

You can use the prime factorization to check whether you have the right number of factors. For example, the prime factorization of 60 is $2 \times 2 \times 3 \times 5$. We can write this using exponents as $2^2 \times 3^1 \times 5^1$. Now we add 1 to each exponent and find the product of these sums: $(2 + 1)(1 + 1)(1 + 1) = 12$. Thus, 60 has 12 factors (including 1 and 60). What is the prime factorization of 156?_____

How many factors does 156 have?_____

Did you find them all above?_____

EXTENSION! How can you use the divisibility rules you studied earlier to find all the factors of a number? Use your method to find all the factors of 184.

	2	3	4	5	6	7	8	9	10
11	12	13	14	15	16	17	18	19	20
21	22	23	24	25	26	27	28	29	30
31	32	33	34	35	36	37	38	39	40
41	42	43	44	45	46	47	48	49	50
51	52	53	54	55	56	57	58	59	60
61	62	63	64	65	66	67	68	69	70
71	72	73	74	75	76	77	78	79	80
81	82	83	84	85	86	87	88	89	90
91	92	93	94	95	96	97	98	99	100

Teacher's Notes for Prime After Prime

No study of number theory is complete without a discussion of prime and composite numbers. In addition to finding prime numbers, this activity presents a unique method for finding all the factors of a number less than 100. Students should complete the activity "Divisibiliti" before attempting this one and should be familiar with the terms prime number, composite number, factor, and multiple. A chart containing numbers from 2 to 100 is given in reproducible form on the following page. The completed factorization chart is also included. Ask students to show their work.

				NCTM Standards					
1	2	3	4	5	6	7	8	9	10
•	•				•			•	

Presenting the Activity

The method for finding all the prime numbers less than 100 was first used by the Greek mathematician Eratosthenes. You are probably familiar with this method, called the Sieve of Eratosthenes. By completing the chart as described on the student page, students can find the factors of the composite numbers less than 100 as well as the prime numbers less than 100. It will take students a fairly long time to fill in the chart, so you may wish to assign that part of the activity as homework on the day preceding your discussion of the activity. As an alternative, you may wish to hand out the completed chart to students and discuss how it was filled in.

After students have a completed chart, discuss some of the information about prime and composite numbers. Ask students to find the composite numbers that have an odd number of factors (4, 9, 16, 25, 36, 49, 64, 81, and 100). These, of course, are the squares of the whole numbers 2 through 10. After a moment's thought, students should see that these numbers have an odd number of factors because one of the factors must be multiplied by itself to get the composite number.

Students can find all the factors of any of the composite numbers simply by looking at the chart and including 1 and the number itself.

Now discuss the prime factorization of a number. Point out that every composite number can be written as a product of prime numbers. To be sure students understand how to use the chart to find the prime factorization of a number, present the following factor trees that use other factors of 24 in the first row.

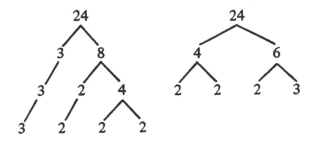

29

The prime factorization of 72 is 2 × 2 × 2 × 3 × 3. One factor tree is

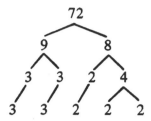

Finding all the factors of a number greater than 100 will be difficult for many students. First, they must use divisibility rules to find a factor less than 100. Then, they can use the chart to find the factors of this number. However, this will not give them all the factors—they must multiply the factors of the smaller number (the number less than 100) by the number they divided by to get the number less than 100. Use 156 as an example. 156 is divisible by 2 and $156 \div 2 = 78$. The factors of 78 are 2, 3, 6, 13, 26, and 39. Each of these factors must be multiplied by 2, and these products must also be included as factors. Thus, the factors of 156 and 1, 2, 3, 4, 6, 12, 13, 26, 39, 52, 78, and 156. (Note that some duplicate factors may occur: $2 \times 3 = 6$, and 6 is already listed as a factor. These duplicate factors are only included once.) Students will probably need additional practice in this method, but first have them complete the next questions on the student page.

The prime factorization of 156 is $2 \times 2 \times 3 \times 13$ or $2^2 \times 3^1 \times 13^1$. Thus, 156 has

$$(2+1)(1+1)(1+1) = 12$$

factors. Using this method gives a good check for students to see if they have found all the factors.

Now ask them to find all the factors of 132 (1, 2, 3, 4, 6, 11, 22, 33, 44, 66, and 132) and of 222 (1, 2, 3, 6, 37, 74, 111, and 222).

If time permits, you may want to discuss other properties of prime numbers. Students may be interested in Goldbach's conjecture, which has never been proved true or not true: Every even number greater than 2 can be written as the sum of two prime numbers. Related to this is that every odd number greater than 7 can be written as the sum of three prime numbers.

Extension

To use divisibility rules to find all the factors of a number, students must check each number. For example, 184 is divisible by 2, so two factors are 2 and 92. It is also divisible by 4, producing the factors 4 and 46. However, this method can become tedious. An easier way is to find the prime factorization of 184 and then find all the possible products of the prime numbers in the factorization. The prime factorization of 184 is $2 \times 2 \times 2 \times 23$. All possible products of these factors gives all the factors of 184: 1, 2, 4, 8, 23, 46, 92, and 184.

	p **2**	p **3**	2 **4**	p **5**	2,3 **6**	p **7**	2,4 **8**	3 **9**
								2,5 **10**
p **11**	2,3,4,6 **12**	p **13**	2,7 **14**	3,5 **15**	2,4,8 **16**	p **17**	2,3,6,9 **18**	p **19**
								2,4,5,10 **20**
3,7 **21**	2,11 **22**	p **23**	2,3,4,6,8,12 **24**	5 **25**	2,13 **26**	3,9 **27**	2,4,7,14 **28**	p **29**
								2,3,5,6,10,15 **30**
p **31**	2,4,8,16 **32**	3,11 **33**	2,17 **34**	5,7 **35**	2,3,4,6,9,12,18 **36**	p **37**	2,19 **38**	3,13 **39**
								2,4,5,8,10,20 **40**
p **41**	2,3,6,7,14,21 **42**	p **43**	2,4,11,22 **44**	3,5,9,15 **45**	2,23 **46**	p **47**	2,3,4,6,8,12,16,24 **48**	7 **49**
								2,5,10,25 **50**
3,17 **51**	2,4,13,26 **52**	p **53**	2,3,6,9,18,27 **54**	5,11 **55**	2,4,7,8,14,28 **56**	3,19 **57**	2,29 **58**	p **59**
								2,3,4,5,6,10,12,15,20,30 **60**
p **61**	2,31 **62**	3,7,9,21 **63**	2,4,8,16,32 **64**	5,13 **65**	2,3,6,11,22,33 **66**	p **67**	2,4,17,34 **68**	3,23 **69**
								2,5,7,10,14,35 **70**
p **71**	2,3,4,6,8,9,12,18,24,36 **72**	p **73**	2,37 **74**	3,5,15,25 **75**	2,4,19,38 **76**	7,11 **77**	2,3,6,13,26,39 **78**	p **79**
								2,4,5,8,10,16,20,40 **80**
3,9,27 **81**	2,41 **82**	p **83**	2,3,4,6,7,12,14,21,27,42 **84**	5,17 **85**	2,43 **86**	3,29 **87**	2,4,8,11,22,44 **88**	p **89**
								2,3,5,6,9,10,15,18,30,45 **90**
7,13 **91**	2,4,23,46 **92**	3,31 **93**	2,47 **94**	5,19 **95**	2,3,4,6,8,12,16,24,32,48 **96**	p **97**	2,7,14,49 **98**	3,9,11,33 **99**
								2,4,5,10,20,25,50 **100**

CHAPTER 2

Geometry and Topology

- Geometric Dissections
- How Many Colors? ♦
- To Stretch a Point
- You Can't Get There From Here
- The Moebius Strip

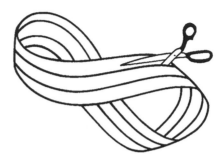

In a standard secondary school mathematics curriculum, geometric and algebraic topics are usually presented separately. Thus, there is little or no overlap between the teaching of these subjects. Pre-algebra is the last course in which both geometric and algebraic operations may be presented in one course, so this section is particularly important.

One of the goals of this series is to acquaint students with topics that are both mathematically significant and not usually presented in a standard course. Most students' future study in geometry will be in Euclidean geometry—to the extent that the majority believe this is the *only* geometry. However, topology logically precedes both Euclidean and non-Euclidean geometries in that these other geometries are special cases of topology. Therefore, four of the activities in this section are addressed to topology problems. The exception is the first activity, "Geometric Dissections."

Many junior high school or middle school basal texts simply ask students to memorize the formulas for areas of various plane figures. "Geometric Dissections" goes a step beyond this and shows why the formulas are as they are. To do this, the activity uses clear, easily perceived transpositions of equivalent areas. Such graphic presentations are not only clear, but most students find the manipulation of these shapes fun, and the last two problems on the student page pose a real challenge to students' ingenuity. This activity is a must.

The classic problem, "How Many Colors?," requires mathematical knowledge only when a proof is attempted. (Such a proof was made possible only recently by using computers.) No proof is offered here and, thus, *all* students can enjoy this activity. This doesn't mean better students will be bored by it—the ideas presented are so different from what they are accustomed to, they are likely to be intrigued. It can be presented any time during the year as an excellent change of pace.

"To Stretch a Point" presents two basic topics in topology—equivalence and the Jordan curve theorem. Although these ideas are not difficult, they require students to think differently mathematically. Don't be surprised if some of your students who make computational mistakes and can't follow formulas understand these topological ideas more easily than some of your better students. Topology depends more heavily on spatial relationships.

Mathematical modeling is used extensively in many applications, and all your potential engineers and scientists will come across it again and again. "You Can't Get There From Here" shows students how to change a physical problem to a mathematical one so it can be solved more easily. Students must also gather data and draw conclusions and generalizations from the data. This, of course, is the basis of scientific experimentation. Although this is a very sophisticated concept, networks are such a simple medium of presentation that even slower students can understand the activity.

"The Moebius Strip" is a fascinating topic and you may want to spend two class periods to thoroughly explore this especially interesting topological figure. It's a good idea to have students work in pairs to reduce the amount of materials necessary and to check each others' results (and, as mentioned in the Teacher's Notes, one maneuver requires three hands). The activity also provides a good exercise in deriving a generalization from empirical observations.

Geometric Dissections

Unlike Humpty Dumpty, geometric figures that have been cut up can be put back together. Moreover, the pieces can be moved around to form different geometric figures.

Look at isosceles trapezoid *ABCD*:

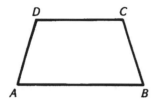

Can you cut off a piece and move that piece to form a parallelogram? Describe where you would make the cut and where you would move the piece. Label any new corners you form._____

_____

Now make one cut on the parallelogram that you made so that you can form a rectangle. Again show your cut and label any new corners.

_____

_____

Make two cuts to divide the triangle

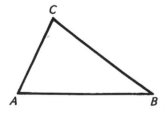

into three pieces. Then use these pieces to form a rectangle. What does this tell you about the formula for the area of a triangle?_____

_____

Use a separate sheet of paper to cut a square into pieces as shown in the accompanying diagram:

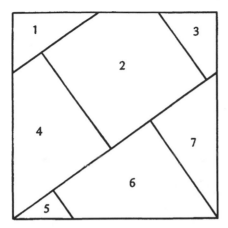

Then using all the pieces, form (1) a rectangle, (2) a parallelogram, (3) three squares of the same size, and (4) an isosceles trapezoid.

EXTENSION! Cut two small hexagons such as those shown in the accompanying diagram. Add the pieces to the third small hexagon to form a large hexagon such as shown at the right of the diagram. If the area of one of the small hexagons is A, what is the area of the large hexagon? If the sides of the small hexagons are one unit long, how long are the sides of the large hexagon?

Teacher's Notes for Geometric Dissections

This activity introduces several geometric transformations and can provide much recreational as well as mathematical value. Perhaps most importantly, the activity can be used to demonstrate the formulas for the areas of triangles, parallelograms, and trapezoids tangibly in terms of their equivalent rectangles. As prerequisites, students should be familiar with the figures discussed in the activity and have been exposed to their area formulas. Ask student to show their work.

					NCTM Standards				
1	2	3	4	5	6	7	8	9	10
	•	•	•		•	•			

Presenting the Activity

It's usually wise to review the definitions of these figures at the very outset of the activity. Students can't very well convert one figure to another if they aren't sure what the result is supposed to look like.

Two other observations should be made regarding this activity. First, the aptitude for this activity is more dependent on people's spatial relations than on their computational abilities. Don't be too surprised if some of your poorer performers outshine some of your normally better students (and this is a good thing). Second, solving these puzzles requires a little bit of imagination. A fair percentage of your class may appear stumped when these questions are first posed. However, when the transformations are presented, they're immediately clear to everyone. In the process, a very convincing demonstration of area equivalency has been given.

The trapezoid can be transformed to a parallelogram by first finding the midpoint of $\overline{BC}$ and calling it E. Then draw a line from E intersecting $\overline{AB}$ at X so that $\overline{EX}$ is parallel to $\overline{AD}$. (This is easy to find because $\overline{CX}$ is perpendicular to $\overline{AB}$.) Now rotate $\triangle XBE$ about E so that B meets C; X will have moved to the point we've designated Y, thus forming parallelogram $AXYD$:

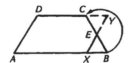 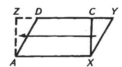

To transform $AXYD$ to a rectangle, drop a perpendicular from C to X. Now translate $\triangle XYC$ to the left until X meets A; Y will then be at D and C will be at Z, forming rectangle $AXCZ$ as shown in the preceding diagram.

For the triangle, students should first locate D and E, the midpoints of $\overline{AC}$ and $\overline{BC}$. (Note that $\overline{DE}$ is parallel to $\overline{AB}$.) Then draw $\overline{CX}$ perpendicular to $\overline{CE}$, forming $\triangle DXE$ and $\triangle EXC$. As before, rotate the triangles about D and E to form rectangle $ABYZ$:

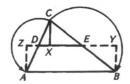

36

Now most students can readily grasp the link between equivalent areas and the area formulas. Because all the pieces of $\triangle ABC$ have been used to form rectangle $ABYZ$, their areas must be the same. The area of $ABYZ$ is given by the base (AB) times either AZ or YB. However, because AZ and YB are each equal to CX, AZ and YB each equal one-half the altitude of $\triangle ABC$. Thus the area of a triangle is *one-half* the base times the altitude.

By similar inspection, students will see that the altitude of both the trapezoid and the parallelogram is CX, not XY or AD. Students will also see that the average of the bases of the trapezoid is shown by AX. In other words, $AX = \frac{AB+DC}{2}$.

Students can trace the figure given on the student page. However, inaccurate tracing or cutting can make this problem a difficult chore, so a reproductible enlargement of this figure (and of the hexagons in the Extension) is provided on the next page. Several solutions are possible; a few are

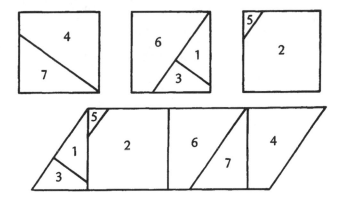

For the rectangle, either butt the three squares together or move piece 4 to the left side of the parallelogram. For the trapezoid, simply flip piece 4 over so that its smallest side is on top.

Extension

Many students will have a tough time coming up with the solution to the Extension. Allow the students enough time to give the problem their best shots, but not so much time that they become overly frustrated and, thence, apathetic.

You can then determine that a side of the larger hexagon is $\sqrt{3}$ times a side of one of the smaller hexagons. Because the area of the new hexagon is three times the area of each of the smaller hexagons, we have verified a significant relationship that holds between similar figures: the ratio of similar figures' areas is the square of the ratio between any two corresponding sides.

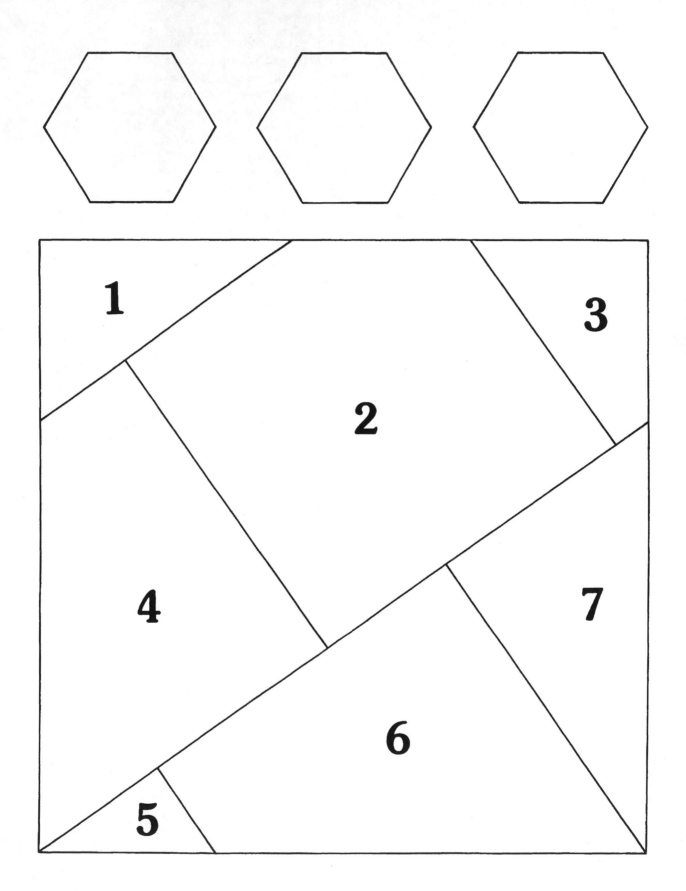

How Many Colors?

Suppose you have a social studies assignment due tomorrow. One part of the assignment is to color a map of South America. To color the map correctly, two countries that have a common boundary must be different colors. If two countries meet only in a point, they can be the same color. You have only four different color marking pens—red, blue, yellow, and green. Will you be able to color the map correctly?_____ On the accompanying map use the letters R, B, Y, and G to indicate how you would color the map.

Suppose you had only two different colors. What three adjacent countries (countries that are next to each other) could you color correctly?_____

_____

_____

What four adjacent countries could you color correctly using only two colors?

_____

_____

Now consider using three different colors. Are there three adjacent countries where you would have to use three different colors? What are they?_____

_____

What four adjacent countries require you to use three colors?_____

_____

What four adjacent countries need four different colors?_____

_____

Do you think you would ever need more than four colors to color a map?_____
Show an example that requires five colors.

EXTENSION! Draw a map that has five regions and requires only two colors to be colored correctly. Then draw a map with five regions that needs three colors. Finally, draw a map with five regions that needs four colors.

Teacher's Notes for How Many Colors?

At first glance this activity may seem to be related more to geography than to mathematics. However, it presents one of the classic problems of topology. Most students will be intrigued by this problem and surprised by its solution. The activity leads students from trial-and-error solving to an analytical method of solution. There are no mathematical prerequisites. Ask students to show their work.

——————————— NCTM Standards ———————————

1	2	3	4	5	6	7	8	9	10
	•	•			•		•	•	•

Presenting the Activity

Students should begin the activity immediately with no preliminary discussion. After they have read the first paragraph, emphasize the rules for coloring the map correctly using the figures

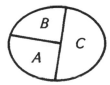 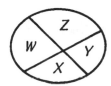

In the above figure at the left, three colors are needed. However, in the above figure at the right only two colors are needed—regions Z and X meet in a point and can be the same color, as can regions W and Y.

Of course, all students will not color the map exactly the same, but all should find that four colors are sufficient.

The next series of questions leads students to an analysis of how many colors are necessary in different situations. To color three adjacent countries using only two colors, the countries must be arranged in one of the following ways

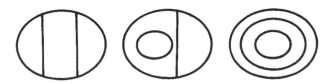

There are several areas of the map where three countries require only two colors; Ecuador, Colombia, and Venezuela or Peru, Bolivia, and Argentina are two examples. Have students present and verify other examples.

41

Ask students how four adjacent countries could be arranged so they would require only two colors. Again, there are several possibilities:

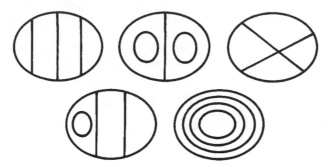

Argentina, Bolivia, Peru, and Ecuador comprise one possibility from the map.

Now consider when three colors are required for three countries. In this case each country must border on the other two. Colombia, Venezuela, and Brazil comprise one example. For four countries to require three colors, exactly two of the countries must not share a common border. Venezuela, Peru, Brazil, and Ecuador comprise one possibility. Again, have students present and verify other examples.

There is only one area of the map where four countries require four colors. Students should realize that in this case, each country must have a common border with the other three. The only four countries on the map for which this is true are Bolivia, Brazil, Paraguay, and Argentina.

Allow students some time to consider the question that immediately precedes the Extension. By now they should realize that to require five colors, each of five countries must border on the other four. Have them spend only a short time on this problem before telling them it's impossible.

Now discuss the history of the four color problem. The problem of *proving* that only four colors are necessary was first proposed in 1840. Until 1977, it was considered one of the famous unsolved problems of mathematics. With the extensive aid of computers, a solution was finally found. This solution was presented by K. Appel and W. Haken in *Scientific American* (vol. 237, No. 4, pp. 108–121, Dec. 1997),

Extension

The following figures are examples using five regions. Other solutions are possible.

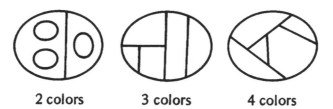

2 colors   3 colors   4 colors

To Stretch a Point

Topology is sometimes called *rubber sheet geometry* or the *mathematics of distortion*. Many of the familiar properties of figures—such as angle measure, length, perimeter, and area—don't apply in topology. For example, among the following figures, a, b, and d are *topologically equivalent*.

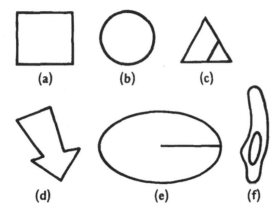

That is, suppose the figures are made of rubber bands. Then each of them can be bent or stretched into each of the others. Figures c, e, and f are not equivalent to a, b, and d because they have different numbers of regions or different numbers of points of intersection. Which of the following figures 1–10 are topologically equivalent?_____

_____

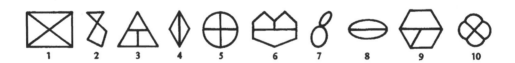

There are certain properties of figures that remain unchanged regardless of how the figure is bent or stretched. Look at the two topologically equivalent figures.

What is true about the points on the figures?_____

_____

These two figures are *simple closed curves*. They divide the plane into exactly two regions—the inside of the curve and the outside of the curve. For these simple closed curves, it is easy to tell whether a point is inside or outside the curve.

In the above figure, is point *A* inside or outside the simple closed curve?_____
How about point *B*?_____
There is an easy way to tell whether a point is inside or outside a simple closed curve. Draw a straight line from point A to the outside of the curve. How many times does the line cross the curve?_____
Now draw a straight line from point *B* to the outside of the curve. How many times does this line cross the curve?_____ Use your results to write a general rule for determining whether a point is inside or outside a simple closed curve._____

_____

EXTENSION! Geometric figures in space can also be topologically equivalent. Consider a cube, open cylinder, sphere, closed cylinder, pyramid, and doughnut. Which of the figures are topologically equivalent?

Teacher's Notes for To Stretch a Point

Topology is a relatively recent branch of mathematics; it was not widely studied until the mid-nineteenth century. Most students will never have heard of it, because it is not usually presented except as a college subject. The basic ideas of topology are not at all difficult and many students will be quite interested in a geometry that does not concern itself with measurement and perfect figures, but rather with the imperfect figures of the real world. Ask Students to show their work.

―――――――――――――――――――――――――――― NCTM Standards ――――――――――――――――――――――――――――

1	2	3	4	5	6	7	8	9	10
		•							

Presenting the Activity

Because most students will be unfamiliar with the ideas studied in topology, Figures (a) through (f) should be thoroughly examined and discussed. Also, be sure students understand what "topologically equivalent" means. If possible, students should work in small groups and use rubber bands to help them determine which of Figures 1–10 are topologically equivalent. Figures 1, 5, and 10 are equivalent; Figures 2 and 7 are equivalent; Figures 3, 6, and 9 are equivalent; and Figures 4 and 8 are equivalent.

Have students look at both the number of regions and the number of points of intersection in Figures 1–10. Although Figures 7 and 8 both form two regions, in Figure 7 there is only one point of intersection, whereas in Figure 8 there are two. Thus, Figure 7 could not be distorted into Figure 8 and the two figures are not topologically equivalent.

Number of regions and number of points of intersection are topological properties because they do not change when a figure is stretched or bent. Another topological property of figures is discussed in the next paragraph on the student page. The *order* of the points on two topologically equivalent figures does not change.

Note that despite the title of this activity, the actual points are not stretched. The distance between points may vary, but their position with respect to one another stays the same. Thus, for the figures shown, point *B* remains between points *A* and *C* in both, even though the distances between *A*, *B*, and *C* have changed.

The rest of the student page before the Extension concerns one of the basic ideas in topology—the Jordan curve theorem. This theorem says that a simple closed curve divides the plane into two regions—an inside and an outside. Although this theorem is intuitively obvious, its proof is quite complicated, particularly for curves such as the last one on the student page.

Students should be familiar with simple closed curves from their study of geometry. However, they have probably not considered simple closed curves like the one containing *A* and *B*.

Begin by explaining that if a point is outside the curve, you can find a path to (say) the edge of the page without crossing a line. If the point is inside the curve, you can't. Point *A* is inside the curve and point *B* is outside.

The number of times straight lines from *A* or *B* to the outside of the curve cross the curve will vary depending on the direction the lines are drawn. However, the line from *A*, an inside point, will cross the curve an *odd* number of times, while the one from *B*, an outside point, will cross an even number of times. The property is obvious if you consider an uncomplicated figure such as a square or circle. You may want to present other figures, such as the following:

Extension

Some students may have difficulty visualizing space (three-dimensional) figures, and it may be helpful to have models of the figures available. The cube, sphere, closed cylinder, and pyramid are topologically equivalent. The doughnut and open cylinder are equivalent if the doughnut is solid and the walls of the cylinder are considered to have thickness. If the doughnut is like an inner tube, the figures are not equivalent.

You Can't Get There From Here

Try to trace Figure 1 without missing any segment and without going over any segment twice.

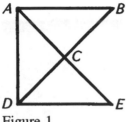

Figure 1

Can you do it? If so, in what order do you pass through the points A, B, C, D, and E? _____

Now try Figures 2–5. Which of these figures can be traced without missing any part and without going over any part twice? Write the order of the points for your routes.

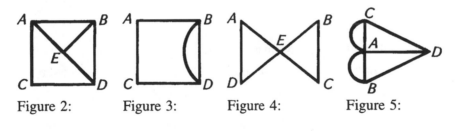

Figure 2:   Figure 3:   Figure 4:   Figure 5:

_____   _____   _____   _____

Each of Figure 1–5 is made up of points (or vertices) that are joined by line segments or arcs. These figures are called *networks*. There can be either an odd or an even number of arcs or line segments that meet at a particular vertex. In Figure 1, for example, A is an odd vertex because three line segments meet there. B is an even vertex because it is where two line segments meet.

Complete the following table for the networks in Figures 1–5:

Figure	Number of Even Vertices	Number of Odd Vertices	Can the Network Be Traced?
1			
2			
3			
4			
5			

How many odd vertices are there in the networks that can be traced?_____

Draw some other networks. Can the networks with zero or two odd vertices always be traced?_____

The Königsberg bridge problem is one of the most famous network problems. The small city of Königsberg is located where a river forms two branches. The people of Königsberg wondered whether a person could walk over each of the city's seven bridges exactly once in a continuous walk through the city. The following figure shows the city and its bridges.

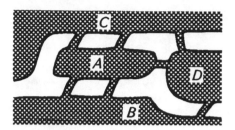

By labeling the regions of land *A*, *B*, *C*, and *D*, and representing the bridges with lines, we can draw the network.

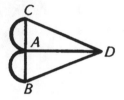

Can the people of Königsberg take the described walk? Why or why not?_____

Suppose another bridge were built between *B* and *D*. Draw the network for this situation. Can the walk be taken now? Why or why not?_____

EXTENSION! Another network problem is the five-room-house problem. Consider the diagram of a five-room house:

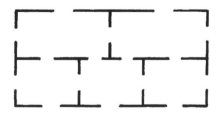

Each room has a doorway to each adjacent room and a doorway leading outside the house. Can a person start either inside or outside the house and walk through each doorway exactly once? Draw a network to find the answer.

Teacher's Notes for You Can't Get There From Here

This activity provides a good lesson in introductory topology. More importantly, it is an excellent tool for developing students' problem-solving skills. The students are presented specific, graphic situations, asked to extract data from them, and then to draw conclusions and generalizations from that data. An acquaintance with the terms vertex, line segment, and arc is the only prerequisite for studying this activity. Ask students to show their work.

————————————————— NCTM Standards —————————————————

1	2	3	4	5	6	7	8	9	10
		•			•			•	

Presenting the Activity

Students should begin immediately with the first problem on the student page. They should find that Figure 1 can be traced without missing any segment and without going over any segment twice. Some possible routes are *A-B-D-A-E-D*, *A-D-E-A-B-D*, and *D-A-B-D-E-A*. There are other possibilities. Be sure all routes found by the students are verified.

Figure 2 cannot be traced. Figure 3 can be traced; some possible routes are *B-A-C-D-B-D*, *D-B-A-C-D-B*, and *D-C-A-B-D-B*. Some possible routes for Figure 4 are *A-D-B-C-A*, *C-B-D-A-C*, and *B-C-A-D-B*. Figure 5 cannot be traced.

The completed table for the networks in Figures 1–5 is

Figure	Number of Even Vertices	Number of Odd Vertices	Can the Network Be Traced?
1	3	2	Yes
2	1	4	No
3	2	2	Yes
4	5	0	Yes
5	0	4	No

Note that there are always an even number of odd degree vertices in a connected network. Networks that can be traced have either zero or two odd vertices. The reason for this is that on a continuous path, the inside vertices must be passed through. That is, if a line "enters" the point, another must "leave" the point. The only vertices that do not conform to this rule are the beginning and end points in the tracing. These two points may be odd vertices. Because there must be an even number of odd vertices, there can only be two or zero odd vertices to trace a network.

In the Königsberg* bridge problem, students take a physical situation and change it to a mathematical model. You may wish to have students experiment with the figure that shows the city and its bridges before considering the network drawn from the physical situation. The network is the same as Figure 5 on the student page. It has four odd

———————

* Today: Kaliningrad, Russia.

49

vertices and, therefore, cannot be traced. The famous Swiss mathematician Leonhard Euler considered the Königsberg bridge problem. In 1735, he proved the walk could not be performed.

If another bridge is built between *B* and *D*, the network is:

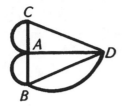

There are now exactly two odd vertices, and the network can be traced. Note that if the eighth bridge were built between *A* and *B*, or *A* and *C*, or *A* and *D*, or *C* and *D*, or *B* and *C*, the network could be traced.

Extension

The accompanying figure shows the network for the five-room house problem.

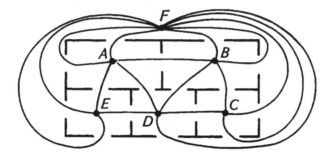

Because there are four odd vertices, the network cannot be traced. Thus, the five-room house problem does not have a solution path.

You may wish to have students devise other similar problems to present to the class.

The Moebius Strip

Topology deals with many different things—from pretzels and knots to networks and maps. The Moebius strip

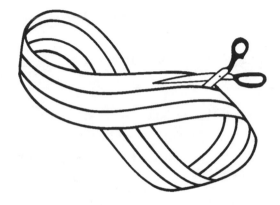

is a favorite of topologists. How many sides does it have?_____ How many edges?_____

You can check your answers by making a Moebius strip. Cut a strip of paper about 10 in. long and 2 in. wide. Make a loop, turn one end over (called a *half-twist*), and tape the ends together. Draw a line in the middle of the strip running lengthwise, continuing until you reach your starting point. Your line never crossed over the edge, did it? Thus, a Moebius strip has only *one* side and *one* edge.

Let's explore some other properties of Moebius strips. Cut the strip you made along the line you drew. What do you get?_____

How many sides does it have?_____ Is it a Moebius strip?_____

How many half-twists does it have?_____

Make another Moebius strip. Cut it parallel to the edge one-third of the way in. Keep cutting until you get back to where you started, always keeping the same distance from the edge. Describe your result._____

_____

Now make another Moebius strip and cut it the same way, only this time, one-fourth of the way in. Describe your result._____

What would you expect to get if you cut one-sixth of the way in?_____

_____

Now let's examine strips with different numbers of half-twists. Make strips with the number of half-twists listed in the left-hand column in the accompanying table. Then complete the second column. Cut each strip down the middle as you did with the first Moebius strip. After you cut each strip, complete the other columns of the table for that strip.

Original Strip		Results After Strip Is Cut		
Number of Half-Twists	Number of Sides	Number of Strips	Number of Half-Twists in Each	Number of Times Separate Strips Are Linked
0				
1				
2				
3				
4				

What general statement can you make from the results in the table?_____

_____

EXTENSION! Make a Moebius strip and draw lines to divide it into thirds lengthwise. Shade the middle third. Cut along the edge of the shaded third. Which loop is shaded? Now see if you can manipulate the loops to form a triple-thick Moebius strip the same length as your original strip. (You will need three hands.) Describe your result.

Teacher's Notes for the Moebius Strip

Many students are fascinated by topology, particularly the Moebius strip. Some of the topological concepts studied in earlier activities are intuitively obvious. The Moebius strip, however, is completely contrary to intuition. Students who find this topic particularly interesting may wish to read articles on the Klein bottle, another one-sided surface. Ask students to show their work.

NCTM Standards

1	2	3	4	5	6	7	8	9	10
		•	•	•	•		•	•	•

Presenting the Activity

Students will need paper, scissors, and tape to construct Moebius strips. It is a good idea to cut about 10 strips for each student before class begins. This allows students more time to work the activity. Graph paper works especially well because it gives students lines to cut along.

Most students will assume that the Moebius strip has two sides and two edges. Even after constructing a strip, they may find it difficult to believe it has only one side and one edge.

When students cut the Moebius strip in half lengthwise, it will not fall apart in two loops as might be expected. Instead it will be a two-sided, two-edged loop (not a Moebius strip), with four half-twists. (*Note*: Tell students to flatten the strip before trying to count the half-twists. Each fold in the strip will be a half-twist.) It is twice as long as the original strip. Of course, the width of the strip will be one-half the width of the original strip.

Cutting a Moebius strip in thirds produces another unexpected result. Note that students must cut past their starting point—cutting around the strip twice. The result is two strips linked together like a chain. One is a Moebius strip the same length as the original, but one-third as wide. The other strip is also one-third as wide as original, but in every other respect is identical to the strip produced if the original Moebius strip were cut in half.

Cutting one-fourth of the way in produces results similar to cutting one-third of the way in. Two linked strips are produced: One is Moebius and the same length as the original; the other is two-sided with four half-twists and twice as long as the original. The only difference is the widths of the strips.

The Moebius strip is two-fourths the width of the original and the other strip is one-fourth the width of the original.

Cutting one-sixth of the way in produces the same two strips as before. The Moebius strip is four-sixths the width of the original and the other strip is one-sixth the width of the original. If students cannot predict this result, have them make Moebius strips and cut them different amounts in from the edge. Some students may be able to generalize their results: A cut $\frac{1}{n}$ from the edge will produce a Moebius strip $\frac{n-2}{n}$ the width of the original and a two-sided strip $\frac{1}{n}$ the width of the original.

53

The table explores the properties of both one- and two-sided strips. The number of half-twists in the original strip influences how many sides it has and also the result when the strip is cut lengthwise. Again, remind students to flatten the strips before counting the half-twists. Unless this is done, it is almost impossible to count the number of half-twists after cutting some of the strips. A completed table given at the end of these notes.

Several generalizations can be made from the table. First, look at strips with an odd number of half-twists. These strips are all one-sided. When they are cut, there is still only one strip. The number of half-twists in this strip is determined by the formula $2n + 2$, where n is the number of half-twists in the original strip.

Then consider the strips with zero or an even number of half-twists. These strips are all two-sided. When they are cut, two separate strips are produced. Each has the same number of half-twists as the original strip and the two strips are linked $\frac{n}{2}$ times. Students should also notice that after cutting, all the strips are two-sided.

Extension

Students will find that the middle, shaded third becomes the smaller, Moebius strip. When they manipulate them to form a triple-thick Moebius strip, they will have a very unusual structure: Two outer "strips" are apparently separated by a Moebius strip "between" them.

Original Strip		Results After Strip Is Cut		
Number of Half-Twists	Number of Sides	Number of Strips	Number of Half-Twists in Each	Number of Times Separate Strips Are Linked
0	2	2	0	0
1	1	1	4	
2	2	2	2	1
3	1	1	8	
4	2	2	4	2

CHAPTER 3

Binary and Exponential Arithmetic

- The Tower of Hanoi
- The Game of Nim
- A Checkerboard Calculator
- The Googol ★

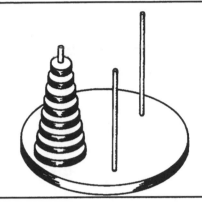

For several years, some teaching formats devoted considerable time to converting numbers among different bases. A lot can be said for the contribution this practice can make to a person's understanding of our number system. However, these conversions are confusing to many teachers as well as students.

Base two, however, deserves considerable attention because it is the foundation of computer operation. As computers become more and more a part of our lives, at least some familiarity with computer language is desirable. We thus shortchange our young people if we don't provide some of the rudiments.

Three activities are included here that provide not only a good introduction to binary mathematics, but contain very engaging and motivating activities. Whereas all of these activities are such fun and contain worthy-to-be-taken-home activities, they are extremely popular and should be presented early in the year.

If its Extension is excluded, "The Tower of Hanoi" is probably the most easily grasped and least time-consuming of the three activities and hence should be presented first. Students who are fascinated by the very large numbers generated in this activity may want to move directly to "The Googol" before reading "The Game of Nim."

During the pilot testing of this program, "The Game of Nim" was one of the most popular activities. Students who enjoy tic-tac-toe will find Nim rewarding. The game lends itself especially well to presentation with an overhead projector, although a chalkboard can be nearly as effective. Practice is important, so it's often a good idea to have students pair off to get the hang of it.

"The Checkerboard Calculator" rounds out the binary activities. In practice, your students will discover that this is not an especially fast or efficient means of calculating. (The advantages of the base-ten system come into play here.) However, the speed of electron flow easily compensates for this disadvantage and students should be aware of the extremely small steps by which calculators and computers operate. This activity also works well in conjunction with "The Googol."

"The Googol" contains a lot of information that students find interesting, yet it doesn't require a great deal of work. Its primary purpose is to introduce scientific notation; thus, you can give many additional examples for practice. A secondary purpose, as stated in the Teacher's Notes, is to discuss the appropriate magnitude of units. This is an area that can easily, yet profitably, occupy two class periods.

The Tower of Hanoi

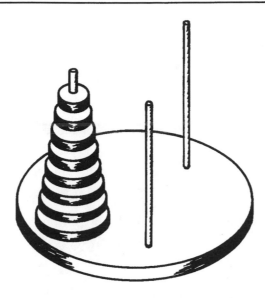

This puzzle was invented in 1883 by the French mathematician Edouard Lucas. Its origin is supposedly based on the following legend:

In the great temple at Benares, beneath the dome that marks the center of the world, rests a brass plate in which are fixed three diamond needles, each a cubit high and as thick as the body of a bee. On one of these needles, at the creation, God placed 64 discs of pure gold, the largest disc resting on the brass plate and the others getting smaller and smaller up to the top one. This is the Tower of Brahma. Day and night unceasingly, the priests transfer the discs from one diamond needle to another, according to the fixed and immutable laws of Brahma, which require that the priest on duty must not move more than one disc at a time and that he must place this disc on a needle so that there is no smaller disc below it. When the 64 discs shall have been thus transferred from the needle on which, at the creation, God placed them, to one of the other needles, tower, temple, and Brahmans alike will crumble into dust and, with a thunderclap, the world will vanish.

How long do you suppose it would take the priests to transfer the discs? Take a wild guess._____

To come up with a more accurate answer—an answer in which you can have some confidence—what two things do you have to know?_____

To get started, transfer a stack of just two discs and use the accompanying table to record the number of moves required. Then try stacks of three, four, and five discs, and again record the number of moves required for each stack. Remember the two basic rules:

1. Move only one disc at a time.
2. Never place a disc on top of a smaller disc.

Number of Discs	Number of Moves
2	
3	
4	
5	

Do you see a relationship—some kind of pattern—between the number of discs and the number of moves needed to transfer the stack? There's a hint given in the box at the bottom of the page, but try to find the relationship before using the hint.

If there are six discs in the stack, how many moves will you need?_____
How about eight discs? (This stack is called the tower of Hanoi.) _____
n discs?_____

How many moves will it take the priests with their 64 discs?_____
If the priests were to make one move every second and work 24 hours a day, 365 days a year, how long will it take them to move the 64 discs?_____

_____

EXTENSION! Is it possible to come up with a table that tells not only how many moves are needed to transfer a stack of discs, but where to move each disc, each step of the way?

Hint: Try adding "1" to each of the "number of moves required."
Does this pattern look familiar?

Teacher's Notes for The Tower of Hanoi

This puzzle provides a dramatic demonstration of the rate at which geometric sequences grow. The Extension problem also gives an intriguing application of binary notation—so very important as electronic data processing increasingly affects our lives. Ask students to show their work.

					NCTM Standards					
1	2	3	4	5	6	7	8	9	10	
	•	•	•	•	•				•	

Presenting the Activity

The Tower of Hanoi, as the puzzle with eight discs is called, is sold commercially, but can be easily constructed by students. An apparatus that closely resembles the commercial version can be made by cutting eight cardboard circles each of different size.

A more easily built set can be made by cutting eight squares of construction paper or cardboard of different sizes and resting them on three plates.

The legend quoted may be at a reading level that may be uncomfortable for some of your students. If so, read this paragraph aloud.

The question, "What two things do you have to know?," is often taken for granted, but because this is a basic problem-solving skill, some attention should be given to it. Pose some additional questions to your class such as, "Fifty percent of the eighth grade attended the basketball game. How many students attended?" Obviously, the question can't be answered unless we know the size of the class. In other words, "Fifty percent of what?" In this case we need to know (a) how many moves and (b) how long per move. For simplicity in calculation, we have assumed the priests can move one disc per second. (This is an optimistic assumption.)

Students should be able to transfer a tower of three discs in seven moves. Now have them try it with four discs. To do this, seven moves are required to transfer the three top discs to one of the other two dowels. This frees the fourth disc which can then be moved to the vacant dowel. Now seven more moves are required to transfer the other three discs back on top of the fourth. Thus, the total number of moves required is 15.

When students consider the game with five discs, they must move the top four discs twice, once to free the bottom disc and once to get them back on the bottom disc after the bottom disc has been moved. Thus, moving five discs takes 31 moves.

Most students now have enough evidence to see that the minimum number of moves is given by $2^n - 1$. Thus, for a stack of eight discs, $2^8 - 1$, or 255, moves are required. For a stack of 64 discs, we'll need $2^{64} - 1$ or 18,446,744,073,709,551,615 moves. At one move per second, this works out to a tad over 580 billion years. Very few students would be able to perform this calculation without guidance. To convert 2^{64} seconds to years we have:

$$2^{64} \div \left(60\frac{\text{sec}}{\text{min}} \times 60\frac{\text{min}}{\text{hr}} \times 24\frac{\text{hr}}{\text{day}} \times 365\frac{\text{day}}{\text{yr}}\right)$$

or 5.8×10^{11} years.

Extension

The Extension problem can be done as independent study or as a class project. It will appeal to the super puzzle solvers and detectives in your class. Have the students number the discs from 1 to 8, smallest to largest. Then have them list the moves for a stack of four discs using binary notation. Next to the binary notation they should describe the action taken. If they're careful, they'll come up with a table such as

Move	Binary	Action Taken
1	1	Move disc 1
2	10	Move disc 2
3	11	Place disc 1 on disc 2
4	100	Move disc 3
5	101	Place disc 1 not on disc 3
6	110	Place disc 2 on disc 3
7	111	Place disc 1 on disc 2
8	1000	Move disc 4
9	1001	Place disc 1 on disc 4
10	1010	Place disc 2 not on disc 4
11	1011	Place disc 1 on disc 2
12	1100	Place disc 3 on disc 4
13	1101	Place disc 1 not on disc 3
14	1110	Place disc 2 on disc 3
15	1111	Place disc 1 on disc 2

To discover which disc to transfer at each move, examine the binary numeral. Count the digits from the right until the first 1 is reached. The number of digits counted tells which disc to move. For example, if the first 1 from the right is the third digit, then the third disc is moved. (See moves 4 and 12.) Looking more closely, students should discover that in all odd-numbered moves (which is all binaries ending with a 1), the smallest disc is moved.

Now the disc's placement must be determined. If there are no other digits to the left of the first 1, then the disc is placed on the dowel that has no discs on it. If there are other digits to the left of the first 1, count digits from the right again until the second 1 is reached. The number of digits counted this time identifies a larger disc that was previously moved. Students must decide whether to place the disc they are moving on top of this larger disc or on the "empty" dowel. To decide which strategy to take, they should count the number of zeros between the first 1 from the right and the second 1 from the right. If there are no zeros between them, or if there is an even number of zeros between them, they should put the disc that they are moving onto the disc that the second 1 refers to. If the number of zeros between them is odd, they should put the disc on the empty dowel.

After examining the table that describes the moves needed to transfer a stack of four discs, your students should be able to construct a table for any number of discs (time permitting) and identify the action taken at any randomly selected move.

The Game of Nim

The game of Nim is believed to have been invented by the Chinese many centuries ago. If you remember your binary arithmetic you can be a master at this fascinating game.

The game is for two players and uses three piles of toothpicks. The players take turns removing toothpicks according to the following rules:

1. In each move, a player may take away toothpicks from only one pile.
2. Each player may take any number of toothpicks, but must take at least one. He or she may take an entire pile at one time.
3. The player who takes away the last toothpick wins.

What does binary arithmetic have to do with this game? Let's begin by reviewing what you know about base-two numbers. Convert the first six numbers in the following chart to base two. Convert the other six numbers to base ten.

Base Ten	Base Two
2	
3	
6	
10	
15	
18	
	100
	111
	1011
	1101
	10011
	10101

Then express base-ten 1 through 15 in base two:

———— , ———— , ———— , ———— , ———— , ———— , ———— , ———— ,

———— , ———— , ———— , ———— , ———— , ———— , ———— .

Now let's see how binary expression can lead to a winning strategy for the game. Two possible combinations of three piles are

Pile #1	14 = 1110																	Pile #1	11 = 1011												
Pile #2	7 = 111										Pile #2	13 = 1101														
Pile #3	13 = 1101																Pile #3	4 = 100					
		2 3 2 2				2 2 1 2

If you arrange the toothpicks as shown, the groups within the piles can be thought of as binary digits. Now add the groups. (This has already been done for you.) If the totals contain one or more odd digits, the three piles are called an *odd set*.

If you can form an even set, you can win the game—*every time*! Let's see how this works. Look at the odd set on the left. What toothpicks can you remove to make it an even set?_____

What toothpicks can you remove to make the set on the right even?_____

Now take a close look at these new sets you've created. Can you remove one or more toothpicks from only one pile and come up with an even set?_____

That's the pickle that you're putting your opponent in. No matter what your opponent does, he or she *has* to form an odd set that you can turn into an even set! If you keep doing this, you eventually wind down to a situation where you present your opponent with something like this: || and ||. What moves can your opponent then make and why does this guarantee that you win?_____

_____

_____

EXTENSION! Suppose you change the rules so that the person who must pick the last toothpick is the loser. What strategy do you take now?

Teacher's Notes for The Game of Nim

This activity has two objectives: First, it provides good practice in using binary numbers via a pragmatic application. Second, it offers a fine demonstration of game strategy—one of the most sophisticated problem-solving skills. Students should be familiar with decimal-to-binary conversions before attempting this activity. Perhaps the greatest beauty of this activity is that it really puts meaningful currency into the student's pockets; students can take home and satisfyingly demonstrate what they've learned from this activity. Ask students to show their work.

————————————————— NCTM Standards ——————————————————

1	2	3	4	5	6	7	8	9	10
•	•			•	•				

Presenting the Activity

Perhaps the most effective (and certainly the most dramatic) way to present this activity is to first bone up on the game. Play a few games with a friend. Then challenge one of your better (perhaps smart-alecky) students to a few games. With the rest of the class watching, play two to four games with one or more willing students. You will win all of the games. Then distribute the student activity sheets and ask the question, "What is it that I know that makes it possible for me to win *every time*?"

By this time your students will have been exposed to the rules of Nim and only a quick review is needed. However, considerable drill is needed to assure that students can quickly and comfortably form binary groupings. Their filled-in table should look like these:

Base Ten	Base Two
2	10
3	11
6	110
10	1010
15	1111
18	10010

Base Ten	Base Two
4	100
7	111
11	1011
13	1101
19	10011
21	10101

Your students can now remove toothpicks so as to form even sets—the winning strategy. Four toothpicks can be removed from any of the piles in the set on the left. The only move that produces an even set on the right is to remove two toothpicks from pile 1. When the students examine these sets, they'll discover that any legal move their opponent makes gives them an odd set that they can turn into an even set.

Now have your students arrange the following sets as shown in the accompanying figure. Set 1: 12, 8, 7;   Set 2: 11, 9, 6;   Set 3: 15, 8, 14;   Set 4: 3, 5, 2.

On the sets shown in the figures, removing three toothpicks from pile 3 of Set 1 works. On Set 2, remove four toothpicks from pile 3. On Set 3, remove nine toothpicks from pile 1. On Set 4, remove four toothpicks from pile 2. These are perhaps best put on the chalkboard. Students can be asked to form even sets and the teacher can easily "remove toothpicks" with an eraser.

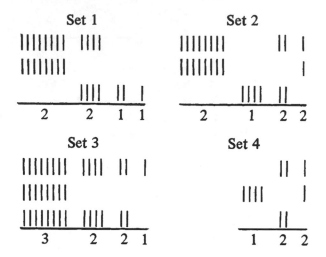

A somewhat nonmathematical, but important, comment should be made here. At first glance, it looks like a very difficult task to convert decimal-system numbers to binary and then add columns to discover even or odd sets. Surely, a player who needs a pencil, tablet, and 5 minutes for each move isn't going to create a very favorable impression on his or her opponent. However, this isn't necessary! If a player arranges the toothpicks as shown, he or she can appear to be merely counting the toothpicks in the piles. Meanwhile, he or she is doing the binary sorting.

Moreover, although we've said that one move cannot convert an even set into another even set, the probabilities are that a random pick from an odd set will produce another odd set. In other words, unless the opponent knows the "secret" of the game, a player can afford to be a little sloppy during the opening moves of the game. After the sizes of the piles are easily manageable, your opponent will probably present you with an easily convertible odd set.

Extension

Players who have mastered the preceding skills will find no difficulty switching to a game in which the last toothpick loses. In fact, a player who wishes to show off a little can let his or her opponent decide halfway through the game whether the last toothpick wins or loses. The winning strategy is still to present your opponent with even sets. A minute's inspection will show that an 0-2-2 set is a winner regardless of whether you're playing last toothpick wins or last toothpick loses. In last toothpick loses, your objective is to force your opponent to draw from either 1-1-1 or 0-0-1.

A Checkerboard Calculator

You probably know that calculators and computers were developed by using binary arithmetic. In this activity you can build a calculator by using a checkerboard. It can be used to add, subtract, multiply, and divide just as a calculator does. There is one difference: Your finger muscles will provide the power instead of an electric current.

Let's begin with an addition problem. Label your checkerboard in binary increments (1, 2, 4, 8, ...) as shown in the figure.

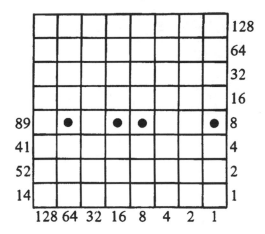

Suppose you're to add $89 + 41 + 52 + 14$. The 89 has been filled in for you: $64 + 16 + 8 + 1$. You fill in the 41, 52, and 14. The 41 is made up of _____ + _____ + _____, the 52 is made up of _____ + _____ + _____, and the 14 is made up of _____ + _____ + _____.

If we now move all of our markers down to the bottom row, the bottom row looks like

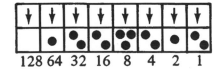

Now we simply regroup from right to left. Two 1s make one 2, so move one marker from the 1's column to the 2's column and set the other one aside. You now have two markers in the 2's column. This makes four, so move one marker to the 4's column and set the other one aside. Continue working to the left in this manner. When you have finished your bottom row should look like

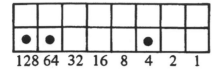

$128 + 64 + 4 = 196$, which is the answer to our $89 + 41 + 52 + 14$. It's really not very fast, but this is how a calculator does it.

The procedure for multiplication is just as simple as addition, and you'll find it goes a lot faster! Suppose you want to multiply 19 times 13. As shown in the accompanying diagram, place markers that add up to 19 just below the checkerboard. Then place the markers that add up to 13 to the right. Now add markers where these rows and columns intersect (as shown by the open circles). All we have to do now is move everything diagonally up and to the right. The column on the right now contains everything we have to add

$$128 + 64 + (2 \times 16) + (2 \times 8) + 4 + 2 + 1 = 247,$$

which is 19×13.

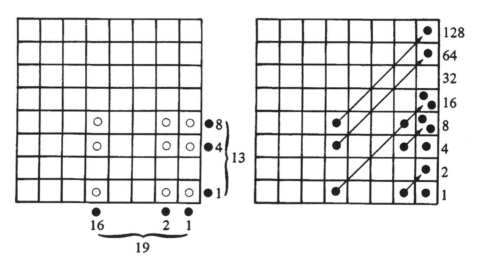

EXTENSION! Now try some subtraction and division problems. The moves you will make are just the reverse of what you did to add and multiply. What do you do with your calculator to find $108 - 83$? $194 - 64$? $250 \div 13$? $361 \div 57$?

Teacher's Notes for A Checkerboard Calculator

This is one of three activities that provide practice in the principles and operations of binary arithmetic. The presentations are somewhat more efficient if the units are given reasonably close together; this eliminates the need to review the basic binary sequence with each topic. Because these activities are so different from each other, there is no worry about students becoming bored with "more of the same old stuff." Indeed, many students are intrigued by the variety of binary applications they discover. For your ease in presenting this activity, a reproducible checkerboard is included on the next page. Ask students to show their work.

NCTM Standards

1	2	3	4	5	6	7	8	9	10
•	•	•	•		•		•	•	

Presenting the Activity

John Napier, the sixteenth-century mathematician who developed logarithms and Napier's bones, also described a method for calculating by moving markers across a chessboard. In addition to being the world's first binary computer, it is also a valuable teaching aid.

The students are first asked to place markers on the board to represent 41, 52, and 14. The markers are positioned by starting at the left and putting a marker on the column of the largest number less than the number to be represented. Then place markers on the next largest number that, when added to the previous number will not exceed the desired total. If the students think of each marker as a 1 and each empty space as a 0, in binary notation 89 is 1011001, 41 is 101001, 52 is 110100, and 14 is 1110.

To add, have students move all markers straight down. Adding the values of these markers will give the correct sum, but to use the board for binary notation, we must first "clear" the row of multiple markers on one cell. Have students start at the right and remove every *pair* of markers on a cell, replacing them with a single marker on the next cell to the left. This will not affect the sum because every two markers with value n are replaced by one marker with value $2n$. In our example, the final result is 11000100_2.

Multiplication is also very simple. As shown on the students' page, a marker is placed on every intersection of a marked column and marked row. Every marker not on the extreme right-hand column is next moved diagonally up and to the right.

Clear the column by halving up as in addition. The desired product is expressed as 11110111_2 or 247_{10}, which students can quickly confirm.

Students will want to know how this works. Markers on the first column keep their values; markers on the second column double in value when moved to the right; markers on the third column quadruple in value; and so on. The procedure can be shown to be equivalent to multiplying with powers to base 2. So our example expresses 19 as $2^4+2^1+2^0$ and 13 as $2^3+2^2+2^0$. Multiplying the trinomials gives $2^7+2^6+2\cdot2^4+2\cdot2^3+2^2+2^1+2^0 = 247$.

Extension

Students can subtract starting at the right and borrowing from cells. Instead, they can alter the entire second row until each marker on the bottom row has one or two markers above it, and no empty cell on the bottom row has more than one marker above it. This can be done by "doubling down" on the second row: Remove a marker and replace it with two markers on the next cell to the right.

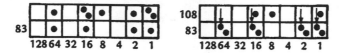

After this, "king" each marker in the bottom row by moving a marker from the cell directly above. The top row now shows the difference of the two numbers in binary notation: $11001_2 = 25_{10}$.

Division is tricky and requires some trial and error. The divisor is marked at the bottom of the board and the dividend on the column at the extreme right. The dividend markers now move down and to the left. This procedure produces a pattern that has markers (one to a cell) only on columns indicated by the divisor. Each marked column must have its markers on the same rows. Only one such pattern can be formed. To do so, it is necessary at times to double down on the right column; that is, remove a single marker and replace it with a pair of markers on the next lower cell. Have students start with the top marker and move it diagonally to the leftmost marked column. Continue with the next marker. If a marker cannot proceed, have students return it to the original cell, double down, and try again. Have them continue in this way, using trial and error to gradually fill in the pattern until the unique solution is achieved.

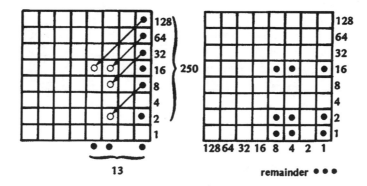

After the final marker is in place, three markers are left over. This represents the remainder (3 or 11_2). The value of the right margin is now 10011_2 or 19_{10}, with $\frac{3}{13}$ left over.

The Googol

10,000,000,000,000,000,000,000,000,000,0...

Before you read any further than this sentence, write the largest number that you can set down using only three digits._____

Did you write 999? Most people do. We can think of a bigger one: 99^9. This would be $99 \times 99 \times 99 \times 99 \times 99 \times 99 \times 99 \times 99 \times 99$, or 913,517,247,483,640,899. 9^{99} would be even larger, wouldn't it? We can come up with a still larger number: 9^{9^9}. Whereas $9^9 = 387,420,489$, 9^{9^9} would be $9^{387,420,489}$. To print this number (using small print) we would need thirty-three 800-page volumes.

You may readily ask, "Isn't it kind of silly to think about numbers this big? Surely they're never used." Well, scientists and economists do use some very, very large numbers. One of the largest numbers that has a name is the *googol*. The name was invented by a child who was asked to think of a name for a very big number, namely, 1 with a hundred zeros after it. A name was then given for a still larger number, *googolplex*. A googolplex is a 1 followed by a googol of zeros.

Even though we've given these numbers names, they're still pretty awkward to handle. So a much easier system has been devised. It is called *scientific notation*. With scientific notation we simply list the first digit on the left, then count the remaining digits and use that number as an exponent of 10. Thus, 6000 is written as 6×10^3 and 400,000 is written as 4×10^5. So you have a number between 1 and 10 up front multiplied by some power of 10.

If more precision is needed, we insert a decimal point. Thus, 6430 is written as 6.43×10^3 and 457,000 is written as 4.57×10^5. Some additional examples follow. Study them and then fill in those that are incomplete:

$$321 = 3.21 \times 10^2; \qquad 1800 = \underline{\hspace{2cm}};$$
$$59,000 = 5.9 \times 10^4; \qquad 100,000 = \underline{\hspace{2cm}};$$
$$4000 = 4 \times 10^3; \qquad 7,600,000 = \underline{\hspace{2cm}};$$
$$6.7 \times 10^2 = 670; \qquad 2.86 \times 10^3 = \underline{\hspace{2cm}};$$
$$1.31 \times 10^6 = 1,310,000; \qquad 8.14 \times 10^7 = \underline{\hspace{2cm}};$$
$$7 \times 10^8 = 700,000,000; \qquad 5.36 \times 10^5 = \underline{\hspace{2cm}}.$$

You can quickly see that scientific notation allows us to write very large numbers much more easily. However, the real usefulness of the system appears when we multiply and divide large numbers. You already know how to multiply a number by 10 or 100 by simply tacking on one or two zeros, and you can divide by 10 or 100 by moving the decimal point one or two places to the left.

What you've been doing is adding powers of 10 to multiply and subtracting powers of 10 to divide. Thus, $240 \times 31,000$ is 2.4×10^2 times 3.1×10^4. We multiply the up front numbers and add the exponents. This gives 7.44×10^6. You try the following problems:

$120 \times 6300 = $ _____; $47,000 \times 2100 = $ _____;

$56 \times 7500 = $ _____ .

To divide $8,400,000$ by 210 we first write $(8.4 \times 10^6) \div (2.1 \times 10^2)$. Divide the up front numbers and subtract the exponents to get 4×10^4. You try these:

$96,000 \div 160 = $ _____; $5,280,000 \div 13,200 = $ _____;

$25,000 \div 6.25 = $ _____ .

EXTENSION! Can you express 0.0046 in scientific notation? Do the preceding rules for multiplication and division apply to such numbers?

Teacher's Notes for the Googol

This activity can be something of an eye-opener for many students. It introduces the extraordinary range of numbers—from very, very small to very, very large. These numbers are not simply theoretical possibilities; they stand for real quantities that are used every day. This, of course, gives an ideal springboard for the study of scientific notation and the ease of its operations. Whether for your own background or as an extra-credit assignment for your students, we strongly recommend Kasner and Newman's article "New Names for Old" (in The World of Mathematics, Vol. 3, pp. 2006–2010). Students should be acquainted with exponents before attempting this activity. Ask students to show their work.

					NCTM Standards				
1	2	3	4	5	6	7	8	9	10
•	•				•		•	•	•

Presenting the Activity

Ask students to answer the first question before passing out the student activity sheets. The temptation to look ahead is just too great.

After reading the first three paragraphs, some students may feel boggled or intimidated. Assure them that they are not alone. No one can visualize a googolplex or nine to the ninth to the ninth.

Most of us can imagine 1000 without much trouble—especially if we visualize a $10 \times 10 \times 10$ cm cube. With some stretching we can visualize 3500, the number of stars that can be seen with the naked eye on a clear night, but an increase of only a few factors of 10 gets most of us out of our depth. A million of anything is really difficult to grasp.

The trick, of course, is to break large numbers into bite-sized chunks. This is why scientists and others use different units for measuring different quantities of the same property (e.g., length). We would not be likely to say that Los Angeles is 28,512,000 in. from San Francisco; we'd say it's 450 miles or a day's drive or an hour's flight.

The mile and the kilometer are convenient units, and plenty big to express distances on earth, but the mile is no better for describing astronomical distances than the inch is for terrestrial distances. The North Star, an easily visible body, is 2.8×10^{14} miles from us. How much easier it is to say 47 light-years. (Because light travels 186,000 miles per second, a light-year is 6×10^{12} miles.)

More important than the convenience of using appropriately sized units is the matter of being able to keep things in perspective. Many well-educated adults have difficulties comprehending theories in evolution and geology because of the enormous amounts of time involved. There seems to be good evidence that the earth is about 5 billion years old. Humans—very primitive humans to be sure—first appeared 500,000 years ago. Now 500,000 is only $\frac{1}{10,000}$ of 5 billion. If we compress the age of the earth to one year, January 1 represents the formation of the earth. If multicellular plants and animals have existed a billion years, they didn't come onto the scene until almost November. Primitive humans then make their appearance around 11:00 p.m. on December 31. All of civilization occurs during the last minute or two of this "geological year," all of science, from Copernicus onward, in the last few seconds!

You may wish to give some additional examples to show your students that very large numbers are not mere mathematical curiosities: We measure our reserves of coal, oil, and natural gas in trillions of tons, barrels, and cubic feet; the GNP of the United States passed the trillion-dollar mark some decades ago; the IRS collected more tax dollars in 2000 than there have been seconds since the birth of Christ. You can probably supply many more examples.

Your students are now aware of the need to manipulate large numbers and, thus, anything that can ease these operations must be considered a boon. You will probably want to walk your students through the problems and suggest additional ones if necessary. The answers to the first six problems on the student page are 1.8×10^3, 1×10^5, 7.6×10^6, 2860, 81,400,000 and 536,000.

The same is true for multiplication and division using scientific notation. When presenting these problems, be sure to have students *first* convert to scientific notation, *then* perform the multiplication or division, and, finally, perform the addition or subtraction of exponents. The answers for multiplication are 7.56×10^5, 9.87×10^7, and 4.2×10^5; the answers for division are 6×10^2, 4×10^2, and 4×10^3.

Now your students are ready to tackle a good, big-number problem such as the following: Suppose your city gets a moderately good soaking and over the past 24 hr it received 2 in. of rain. If your city is a square, 5 mi on a side, how many gallons of water fell on it? First we calculate the cubic inches. A mile is 5280 ft and a f is 12 in., so we have $(5 \times 5.28 \times 10^3 \times 12)^2$ to give us the area of the city in square inches. This is 1.0×10^{11}. We said 2 in. of rain fell, so the volume of rain is 2×10^{11} in.^{3} There are 231 in.^{3} per gallon, so we divide by 0.231×10^3 to get 8.66×10^8 or 866,000,000 gallons—nearly a billion gallons.

Extension

The answer here is, of course, yes, 4.6×10^{-3}. You will probably want to reinforce this idea with the following examples:

$$2.4 \times 10^{-6} \times 4 \times 10^4 = 9.6 \times 10^{-2},$$
$$3.2 \times 10^{-4} \times 1.5 \times 10^{-2} = 4.8 \times 10^{-6},$$
$$8.4 \times 10^8 \div 1.4 \times 10^{-3} = 6 \times 10^{11},$$
$$7.35 \times 10^{-4} \div 1.05 \times 10^{-8} = 7 \times 10^4.$$

CHAPTER 4

Problem-Solving Strategies

- Posers
- The Euclidean Algorithm
- Formulas—Mathematical Shorthand
- Which Formula?
- Writing Formulas ⋆

As mentioned earlier, the problem-solving aspect of the activities cannot be overemphasized. There is a growing concern with what has been termed math illiteracy, and problem solving has been given the top priority in the NCTM's *Principles and Standards for School Mathematics* (2000).

Although all of the activities in this book present complex, multistep problems for the students to work through, the activities in this section address the strategy and methodology of problem solving. Because the concepts treated in this section have such wide applicability to your other work, I recommend you present them as early in the year as is practical.

If there is any unit in this book that I feel is a *must*, it's "Posers." The activity presents a variety of problems—some of them very old chestnuts and all of them amusing. Their common thread is that thinking about what is asked for is more important than the actual calculations required. Beyond that, students can see what happens when they're too quick on the draw to make their assumptions.

"The Euclidean Algorithm" is based on a topic that is treated in most pre-algebra texts, the greatest common divisor or factor. Unfortunately, little is done to show the applications or usefulness of the idea. "The Euclidean Algorithm" presents a feasible, realistic situation and asks students to work out the solution to a problem. They begin by trial and error and then progress to a more systematized and efficient attack on the problem.

Because formulas are so important to an even rudimentary facility in mathematics, and because they're often so poorly understood and misused, we've devoted three activities to the use, selection, and generation of formulas. Taken together, the three activities on formulas offer a more comprehensive treatment of the topic than is now found in any text. Moreover, you'll find the ideas in these units useful to refer to throughout your course—especially whenever word problems are discussed.

74

"Formulas—Mathematical Shorthand" goes to the heart of the matter immediately and asks students to think of formulas as sentences. They then examine what happens when one word of the sentence is changed; then what happens when a different word is changed. Finally, they see the effect on the sentence of changing both words at the same time.

"Which Formula?" considers the next logical step. Too often students plug values into formulas without thinking about whether they're using the appropriate formula. This activity prompts students to ask what information (or quantity) they're trying to discover. They then compare this with the information already available to them and see if one part of the "sentence" is compatible with the other.

"Writing Formulas" rounds out this group of activities. It begins by reviewing and combining the two previous activities and then introduces the concept of a variable. More importantly, students are shown how to create formulas to fit the particular problem at hand when an off-the-shelf, standard formula doesn't exist. The idea of superfluous and redundant information is also introduced.

Posers

An egg carton held 12 eggs when it was full, but now contains 9. Take 3 eggs from the carton. How many eggs do you now have?_____

Because his parents were away last night, Patrick stayed up very late to watch a monster movie and other cultural programs. He's been very tired today and decides he'll go to bed early this evening—at 7 o'clock—and get a good night's sleep. He has to be up for school at 8:30, so he sets his alarm clock for that time. If he falls asleep right away, how much sleep will he get by the time the alarm rings?_____

With most mathematics problems, it's more important to be able to formulate the problem than to perform the calculations. You can always go back and check your arithmetic. But if you haven't asked the right questions in the first place, checking your arithmetic won't help much, will it? The mathematics needed for the preceding two problems is very simple, yet it's amazing how many people will answer without thinking,"6 eggs" and "$13\frac{1}{2}$ hours." Where are they making their mistakes?_____

_____

A lot of problems are solved more easily if you use what is sometimes called *the pigeon-hole principle*. This is nothing more than keeping track of what is happening each step of the way. Suppose you have six pairs of blue socks and four pairs of black socks jumbled together in a drawer. In the dark, how many must you pull out to be sure of getting a matched pair? _____

How many socks must you pull out to be sure of getting a blue pair?_____

A similar kind of problem is the following: An apartment house has 25 mail-boxes. If 51 letters are delivered, proved that *at least* one mail box receives *at least* 3 letters._____

What conclusions can you make if 52 letters are delivered?_____

If 80 letters are delivered?_____

Finally, although this problem has been around even longer than the author, it still stumps a lot of people: Carl gave a hotel clerk $15 for his room for the night. The clerk later discovered that he had overcharged by $5 and sent a bellboy to Carl's room with five $1 bills. The dishonest bellboy gave only $3 to Carl, keeping the other $2 for himself. Carl paid $12 for his room. The bellboy got $2. This accounts for $14. Where is the missing dollar?_____

EXTENSION! Consider an airplane that flies 400 miles from Chicago to Pittsburgh and then 400 miles back to Chicago. On the way to Pittsburgh the plane has a tail wind and averages 240 miles per hour. On the way back it's flying into a head wind and averages 160 mph. What is the plane's average speed for the round trip?

Teacher's Notes for Posers

"Posers" is the first of the problem-solving activities that should be presented. The mathematics of this activity is very simple, but it tackles head-on the idea of thinking about the problem at hand—a skill needed for all of the activities. Ask students to show their work.

NCTM Standards

1	2	3	4	5	6	7	8	9	10
	•			•	•			•	•

Presenting the Activity

The first two problems are answered incorrectly by people who blurt out an answer before thinking about what's been asked. For the first problem "you" now have three eggs; the *carton* now has six eggs. Patrick, of course, gets $1\frac{1}{2}$ hours of sleep before the alarm goes off, not $13\frac{1}{2}$. (Some students may point out that many digital alarms distinguish between a.m. and p.m. Simply indicate that we are using an analog clock.)

There are hundreds and hundreds of posers such as this, ranging from the simple-minded, "Heads I win; tails you lose" and "OK, if I'm wrong, you buy the cokes," to mathematically more subtle ones you may wish to present to the class: A South Pacific island has 300 women among its inhabitants. Five percent of them wear one earring (each). Of the remainder, half wear two earrings and half wear none. Altogether how many earrings are worn? Now if half the women are wearing two earrings and half none, that's the same as all the women wearing one, so the answer is 300 and we see that the 5% information is just so much extraneous dust.

The pigeon-hole principle is essentially a matter of asking what is the effect of each action on the whole problem. Consider what happens with the first sock problem: If, after withdrawing the first sock in the dark, someone turns on the lights, you can place the sock into one of two compartments or pigeon holes—one labeled blue, the other black. When you withdraw a second sock, you find that if either matches the first (i.e., you already have your pair) or you place it in the other compartment. The third sock, then, must match *at least* one of the first two socks pulled. Note that the number of pairs given is again extraneous information. There could be 50 pairs of blue socks and 1 pair of black socks and the answer is still 3.

In the next problem the number of pairs of socks does matter. In the worst imaginable case—highly improbable, but possible—you could pull all 8 black socks before getting a blue one. Therefore, 10 socks must be withdrawn to be *certain* of getting a blue pair. Substitute some other numbers and a third or fourth color to be sure all students understand what's going on here.

Using similar reasoning for the next problems, ask students to put themselves in the place of what, in simpler days, we called the mailman. Students should imagine themselves holding a stack of 51 letters and placing them one at a time in consecutive mailboxes. After running the gamut of the boxes twice one letter remains. Therefore, at least one box must contain at least three letters. (Notice that "at least" is italicized on the student page; an acceptable solution is for one box to contain 51 letters.)

With 52 letters, at least two boxes must contain 3 or more letters *or* at least one box must contain 4 or more letters. With 80 letters, we distribute 3 letters to each box and have 5 letters left over, so one box must contain 8 or more letters, or five boxes must contain 4 or more letters, or the combinations in between.

The last problem is truly a classic poser, so give students ample time to puzzle over it. The statement of $12 plus $2 equals $14 is beside the point and quite misleading. Carl did not pay $12 for the room; he paid $10. The $12 figure is what he paid for the room plus what the bellhop appropriated. *This* $12 plus his $3 change accounts for his $15. Using a number line clearly shows there is no reason for adding $12 and $2.

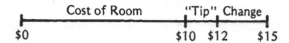

You may then wish to give another version of this problem: Three businessmen check into a hotel and each chip in $10 for the room they will share. The clerk later discovers he has overcharged; the room is supposed to cost $25. He gives the bellhop five $1 bills to take to the men. The three men are unable to figure how to equally divide the $5, so each takes $1 and they give the bellhop $2. The men have each paid $9 for the room. Three times $9 is $27. The bellhop got $2, making $29. Where did the other $1 go?

Extension

The quick (and incorrect) answer is given by $\frac{160+240}{2}$ or 200 mph for the average speed. But the arithmetic mean doesn't work here because the airplane does not fly these speeds for the same amount of time. This becomes clear when we treat it as an ordinary $d = rt$ problem. From Chicago to Pittsburgh the time in hours is $400 \div 240 = 1.667$; from Pittsburgh to Chicago it is $400 \div 160 = 2.5$ hours. The time for the round trip is thus $2.5\,\text{hr} + 1.667\,\text{hr} = 4.167\,\text{hr}$ and the average speed is $800\,\text{mi} \div 4.167\,\text{hr} = 192\,\text{mph}$.

Notice that here too, extraneous information is given: We don't need to know the distance between Chicago and Pittsburgh; we can just call it d. Then, $t_1 = \frac{d}{240}$ and $t_2 = \frac{d}{160}$. The time for the round trip, T is $\frac{d}{240} + \frac{d}{160} = \frac{d}{96}$. The rate for the round trip is then $2d = r \cdot \frac{d}{96}$ and, thus, $r = 192$ no matter what the value of d. You may wish to give different cities and distances to lead students to this conclusion.

The Euclidean Algorithm

Suppose you and several friends have spent the summer in Nevada County panning for gold. The summer is nearly over now and you have a sack of gold dust to distribute among your friends. Some people worked many more days than others, and you've agreed that the people are entitled to the following amounts of gold dust:

A—12 oz   D—4 oz

B—1 oz   E—6 oz

C—3 oz   F—11 oz

You take out your beam balance and set of weights and find that most of the weights are missing. You have several 5-oz and several 7-oz weights, but that is all. Can you use just these weights to measure out the right amount of gold dust for each person? Describe how you would arrange the weights to get:

12 oz_____

1 oz_____

3 oz_____

4 oz_____

6 oz_____

11 oz_____

What if you had been left with a stack of 2-oz and 4-oz weights? Would you still be able to measure the required quantities? What about a set of 3-oz and 9-oz weights? Explain:_____

_____

_____

What you have just learned is an application of the *Euclidean algorithm*. Look at the following sets of weights

Weights	GCD
5,7	_____
2,4	_____
3,9	_____
8,20	_____
5,25	_____

For each of them, write the largest number that can be evenly divided into both numbers of the set. (This number is called the *greatest common divisor*, or GCD.)

The GCD tells you the *smallest* amount that can be weighed with the given set of weights, and anything you weigh must be an integral multiple of the GCD.

At first glance you might think it's difficult to find the GCD of two larger numbers, say, 219 and 945, but it's really quite simple: just keep dividing by remainders until you get a zero remainder. The last nonzero *remainder* is the GCD. For this example, divide 945 by 219. $945 = (4)(219) + 69$. The remaining steps are

$$219 \div 69: \quad 219 = (3)(69) + 12,$$
$$69 \div 12: \quad 69 = (5)(12) + 9,$$
$$12 \div 9: \quad 12 = (1)(9) + 3,$$
$$9 \div 3: \quad 9 = (3)(3) + 0,$$
$$GCD = 3.$$

Now find the GCDs of (a) 12 and 45; (b) 121 and 384.

EXTENSION! Find a way to express the GCD in terms of multiples of the two numbers. In other words, how many 945-oz weights and how many 219-oz weights would be needed to weigh 3 oz?

Teacher's Notes for The Euclidean Algorithm

Most basal texts give some consideration to finding greatest common divisors or factors. However, they too often give no applications, leaving the student to wonder why they have learned this lesson. This activity provides a practical and interesting application and at the same time presents yet another useful problem-solving strategy. Ask students to show their work.

NCTM Standards

1	2	3	4	5	6	7	8	9	10
•	•								

Presenting the Activity

No preliminary discussion is necessary before students begin the activity. To give students an idea of how the weighing is to be done, point out that the scale on the student page shows how 9 oz of gold could be weighed. Some students will approach the problem using a trial-and-error method. Encourage them to develop a more systematic approach by writing multiples of 5 and 7:

> Multiples of 5:   5, 10, 15, 20, 25, 30, ...
>
> Multiples of 7:   7, 14, 21, 28, 35, 42, ...

By comparing multiples, they should quickly see how to weigh the various amounts:

12 oz: one 5-oz and one 7-oz on the same pan, the gold on the other pan

1 oz: three 5-oz on one pan, two 7-oz plus the gold on the other pan

3 oz: two 5-oz on one pan, one 7-oz plus the gold on the other pan

4 oz: two 7-oz on one pan, two 5-oz plus the gold on the other pan

6 oz: three 7-oz on one pan, three 5-oz plus the gold on the other pan; or four 5-oz on one pan and two 7-oz plus the gold on the other pan

11 oz: three 7-oz on one pan, two 5-oz plus the gold on the other pan

When students use only 2-oz and 4-oz weights, they can only measure quantities that are multiples of 2. Similarly, with only 3-oz and 9-oz weights, they can only weigh multiples of 3.

Ask students if they can think of any amount that they *can't* weigh with the 5-oz and 7-oz weights. They won't be able to come up with any such amount. Then have them find the greatest common divisors (sometimes called the greatest common factor) in the table on the student page. The GCDs for the weights given are 1, 2, 3, 4, and 5, respectively. Emphasize that the GCD of the two weights is the *smallest* amount that can be weighed, and the two weights can only weigh multiples of the GCD.

Next, students discover how easy it is to find the GCD of *any* two numbers.

For 12 and 45,

$45 \div 12$:　　$45 = (3)(12) + 9,$
$12 \div 9$:　　$12 = (1)(9) + 3,$
$9 \div 3$:　　$9 = (3)(3) + 0.$

Thus, the GCD is 3.

For 121 and 384,

$384 \div 121$:　　$384 = (3)(121) + 21,$
$121 \div 21$:　　$121 = (5)(21) + 16,$
$21 \div 16$:　　$21 = (1)(16) + 5,$
$16 \div 5$:　　$16 = (3)(5) + 1,$
$5 \div 1$:　　$5 = (5)(1) + 0,$
$GCD = 1.$

If necessary, students can practice this technique with other pairs of numbers. This method of determining the greatest common divisor is much easier than the methods usually taught—listing all the factors of each of two numbers or finding their prime factorizations.

Extension

Be careful: The Extension is tricky. To express the GCD in terms of multiples of the two numbers, substitute various values using the equations used to find the GCD. Suggest first trying to develop a method for doing this with smaller numbers, such as the 12 and 45 in (a) on the student page. The equations used to find the GCD for these two numbers are given in the previous paragraph. From the second equation, we can write $3 = 12 - 9$. From the first equation, $9 = 45 - (3)(12)$. We substitute this value for 9 into

$$3 = 12 - 9$$
$$= 12 - [45 - (3)(12)]$$
$$= (1)(12) - 45 - (3)(12)$$
$$= (4)(12) - (1)(45).$$

Thus, the GCD is written in terms of multiples of the two numbers. It is thus necessary to use four 12-oz weights and one 45-oz weight to weigh 3 oz.

Now this idea can be extended to 219 and 945. The equations from the student page are

$$945 = (4)(219) + 69,$$
$$219 = (3)(69) + 12,$$
$$69 = (5)(12) + 9,$$
$$12 = (1)(9) + 3,$$
$$9 = (3)(3) + 0.$$

We start with the fourth equation and write $3 = 12 - 9$. Whereas $9 = 69 - (5)(12)$,

$$3 = 12 - [69 - (5)(12)] = (6)(12) - 69.$$

From the second equation, $12 = 219 - (3)(69)$, so,

$$3 = (6)[219 - (3)(69)] - 69$$
$$= (6)(219) - (18)(69) - 69$$
$$= (6)(219) - (19)(69).$$

Now, from the first equation, $69 = 945 - (4)(219)$; hence,

$$3 = (6)(219) - (19)[945 - (4)(219)]$$
$$= (6)(219) - (19)(945) + (76)(219)$$
$$= (82)(219) - (19)(945).$$

Thus, 82 219-oz weights and 19 945-oz weights would be needed to weigh 3 oz.

Formulas—Mathematical Shorthand

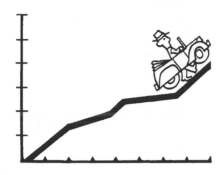

One of the best ways to understand mathematics is to think of it as a language. You'll find that it's a very precise and concise language, which means you can say exactly what you want to say in a very small amount of space.

As with any language, of course, you have to know what the different letters and symbols stand for before the language makes any sense. Then you can translate English sentences to mathematical sentences and use them to solve problems.

A formula uses symbols and letters to show how certain quantities are related. Look at the formula

d stands for distance traveled ⟶ *r* stands for rate of travel (speed)

$$d = r \times t$$

t stands for length of time

You already know what the symbols = and × mean, so $d = r \times t$ is a formula or mathematical sentence. To express it in English words you would write:____

_____

_____

If you know the *rate* or speed at which a vehicle has been traveling and the length of *time* it has been traveling, then you can find the *distance* it has traveled. Suppose a car has been going 40 mph for 4 hr. How far has the car gone?

Table 1

Distance	Rate	Time
	30 mph	1 hr
	30 mph	2 hr
	30 mph	3 hr
	30 mph	4 hr
	30 mph	5 hr

Table 2

Distance	Rate	Time
	10 mph	2 hr
	20 mph	2 hr
	30 mph	2 hr
	40 mph	2 hr
	50 mph	2 hr

Table 3

Distance	Rate	Time
	10 mph	1 hr
	20 mph	2 hr
	30 mph	3 hr
	40 mph	4 hr
	50 mph	5 hr

In Table 1, rate (speed) stays the same and time increases. In Table 2, rate increases and time stays the same. In Table 3, both rate and time increase. Find the distances in the tables.

What happens to the distance when the rate stays the same and the time of travel is increased?_____ Do the distance and time increase the same way? If so, how?_____

What happens to the distance when travel time is held the same, but the rate is increased?_____

Do the distance and rate increase the same way? If so, how?_____

What happens to the distance when both the rate and time increase?_____

Does the distance increase the same way as the time?_____ The same way as the rate?_____ How does the distance increase?_____

Now consider a 200-mi trip. If you travel 50 mph, can you make the trip in 2 hr?_____ If not, how many hours will it take?_____

What speed would you have to go to get there in 2 hr? _____

EXTENSION! The volume of a rectangular box is found using the formula $V = l \times w \times h$, where V stands for the volume, l stands for the length of the box, w stands for its width, and h is the height of the box. What happens to the volume (1) if the length is doubled? (2) if both the length and width are doubled? (3) if the length, width, and height are all doubled?

Teacher's Notes for Formulas—Mathematical Shorthand

This activity goes to the core of many students' most troublesome area—word problems. Their stumbling block is often overcome if they can think of formulas as sentences. (The word equation is purposely avoided.) Ask students to show their work.

					NCTM Standards					
1	2	3	4	5	6	7	8	9	10	
	•		•	•				•	•	

Presenting the Activity

Discuss the idea of mathematics as a language. To solve a word problem, we must first translate the English sentences into mathematical sentences—just as we may translate English into French or German.

In some ways, a mathematical formula is similar to a recipe. To use a formula or recipe, we have to know two things: what the symbols stand for and when the formula applies. We first consider what the symbols stand for in the formula $d = r \times t$: distance traveled equals the rate of travel (speed) multiplied by the length of time traveled. Thus, this formula applies whenever we study uniform motion such as in the train and car problems often found in mathematics books.

After students decide the formula is applicable, they need only substitute given values to find the unknown values. Thus, for the car traveling 40 mph for 4 hr.

$$d = r \times t$$
$$= 40 \times 4$$
$$= 160.$$

The car has traveled 160 mi.

The remainder of the activity leads students through an analysis of the formula by considering how changing one or more value affects other values. The completed tables are

Table 1

Distance	Rate	Time
30 miles	30 mph	1 hr
60 miles	30 mph	2 hr
90 miles	30 mph	3 hr
120 miles	30 mph	4 hr
150 miles	30 mph	5 hr

Table 2

Distance	Rate	Time
20 miles	10 mph	2 hr
40 miles	20 mph	2 hr
60 miles	30 mph	2 hr
80 miles	40 mph	2 hr
100 miles	50 mph	2 hr

Table 3

Distance	Rate	Time
10 miles	10 mph	1 hr
40 miles	20 mph	2 hr
90 miles	30 mph	3 hr
160 miles	40 mph	4 hr
250 miles	50 mph	5 hr

In Table 1, as the time increases, the distance increases at the same rate. That is, if time is doubled, the distance traveled is twice as far. Similarly in Table 2, when the rate is doubled, so is the distance. In Table 3, both the time and rate increasing. When both the time and the rate double, the distance traveled is four times as great.

The results found in the tables can be seen mathematically by using the formula. If $d = r \times t$, multiplying r or t by 2 multiplies d by 2. When both r and t are multiplied by 2, d is multiplied by 4.

Useful as the tables are for helping students to understand formulas, graphs used simultaneously help grease the skids even more. The graph

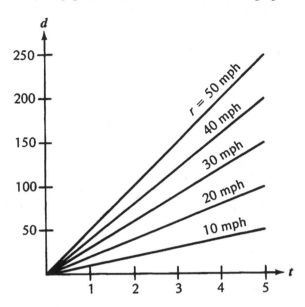

contains all of the information given in the three tables. Note that it uses d as the vertical axis and t as the horizontal axis—though these can be reversed. In either case the five different rs give the slopes of the lines. We strongly recommend that during your discussion you use such a graph, either on the chalkboard or on an overhead projector.

For the 200-mi trip, students should easily see that by traveling 50 mph for 2 hr you will only travel 100 mi. If you want to go twice as far, you must travel twice as long—4 hr. If you want to travel the 200 mi in 2 hr you must travel twice as fast, 100 mph.

Students might like to see another example. Consider $E = r \times t$, where E equals earnings per year, r equals rate of pay per hour, and t equals the number of hours worked. Plug in different values for r and t to see the effect on E.

Extension

Most students should be able to answer the questions in the Extension because they follow directly from the analysis in the tables. If the length doubles, the volume doubles. If the length and width both double, the volume is multiplied by 4. If the length, width and height all double, the volume is multiplied by 8. This can easily be verified using specific values for the length, width, and height.

Which Formula?

Formulas are like recipes—to use them you have to know what the symbols stand for and when you should use a particular formula. You should be familiar with all the following formulas

$$P = 2l + 2w,$$
$$P = 4s,$$
$$C = 2\pi r,$$
$$A = l \times w,$$
$$A = s^2,$$
$$A = \pi r^2.$$

That is, you should know what all the symbols and letters stand for. Do you always know when to use these formulas? Suppose you want to find the area of a rug. Which formula would you use?_____

What do you need to know before you can decide which formula to use?_____

If the rectangular rug is 12 ft long and 8 ft wide, which formula should you use? _____ If the radius of a circular rug is 5 ft, which formula should you use? _____ Sometimes you have to use two formulas to solve a problem: The perimeter of a rectangular rug is 54 ft. Its length is 15 ft. What is the area of the rug? _____

What formula should you use to find the area?_____

What do you need to know before you can use the formula?_____

What other formula can you use to find this information?_____

What is the width of the rug?_____ What is its area?_____

Now let's try a more complicated problem: A rectangular room is 16 ft long and 11 ft wide. A square rug, 10 ft on each side, is on the floor of the room. What is the area of the floor *not* covered by the rug? Begin by drawing a picture. Label all the information you have. What is the formula for finding the area of the rug?

What is this area?_____

What is the formula for finding the area of the floor?_____

What is this area?_____

How can you use this information to answer the question in the problem?_____

_____

What is the area of the floor *not* covered by the rug?_____

There is another way to solve this problem that doesn't use the area of either the rug or the floor. Imagine the rug in one corner of the floor with two of its edges touching two of the walls of the room. Describe how you could find the area of the floor not covered by the rug._____

_____

Now try this problem: A square room has a circular rug on the floor. The rug just touches all four walls. If the circumference of the rug is 12π ft, what is the area of the floor *not* covered by the rug? (Draw a picture and use 3.14 for π.)

EXTENSION! The perimeter of a rectangular room is 44 ft. The length of the room is 12 ft. Will a circular rug with an area of 113.04 ft^2 fit on the floor in the room? (Use 3.14 for π.)

Teacher's Notes for Which Formula?

This activity is the second in a series of activities using formulas to solve problems. In the first activity of the series, "Formulas—Mathematical Shorthand," students were introduced to formulas and use one formula to solve several different problems. In this activity, students are presented with six formulas and must decide which formula applies in a particular problem. Ask students to show their work.

NCTM Standards

1	2	3	4	5	6	7	8	9	10
	•	•	•	•					•

Presenting the Activity

Begin with a general discussion of formulas, reminding students that formulas show how quantities are related. Then discuss in detail each of the six formulas at the top of the student page. Be sure students understand what each letter stands for. Ask students to state in words what each formula means.

Students should realize that they can't decide which formula to use to find the area of a rug until they know the shape of the rug. For a rectangular rug 12 ft long and 8 ft wide, they would use the area formula for a rectangle, $A = l \times w$. For a circular rug with a radius of 5 ft, they should use the area formula for a circle, $A = \pi r^2$.

Students should be encouraged to read each problem carefully. Too often they read a problem once and decide they don't know how to solve it. If they read it again, perhaps even a third time, they will usually understand what's going on. The questions on the student page for the next problem lead students through one analysis of the problem. They should try to think of questions like these each time they begin to solve a word problem. They should use the formula $A = l \times w$, because they know the rug is rectangular because the length is given. Before they can use the area formula, they must know the width of the rug. The perimeter and length of the rug are given, so the width of the rug can be found using the formula, $P = 2l + 2w$. If students have difficulty using this formula to find that the width is 12 ft, work through the problem on the chalkboard using diagrams. After they know the width of the rug, students should easily be able to find the area, 180 ft^2.

A drawing for the next problem is as follows:

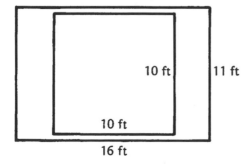

Students should see that subtracting the area of the rug ($A = s^2 = 10 \times 10 = 100$) from the area of the floor ($A = l \times w = 16 \times 11 = 176$) will give them the area of the floor *not* covered by the rug. Thus, $176 - 100 = 76$ ft^2 of floor not covered by the rug.

If the rug is positioned as shown in the diagram

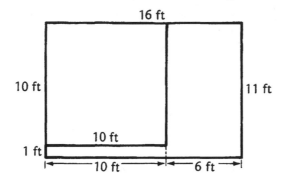

the dimensions of the area not covered by the rug can be found easily. The area of the two rectangles indicated on the figure are then added ($10 \times 1 = 10$; $6 \times 11 = 66$; $10 + 66 = 76$). This problem should help students realize there may be more than one way to solve a problem.

The drawing for the next problem is

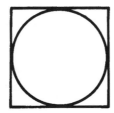

The reasoning is as follows: Because the circumference of the rug is 12π, its radius is 6 and its area is 36π or 113.04 ft^2. Because the rug touches the walls, the length of each side of the room must be the same as the diameter of the rug, that is, 12 ft. Thus, the area of the floor is 144 ft^2 and the area not covered by the rug is $144 - 113.04 = 30.96$ ft^2. The trick to this problem, of course, is to see that the diameter of the rug is the same as the length of each wall.

Extension

This problem is not difficult and should be completed by the entire class. The only added skill is common sense. If students find the area of the floor, they will see that it is greater than the area of the rug, and the rug will appear to fit on the floor. However, the diameter of the rug is 12 ft and the width of the room is only 10 ft. Thus, the rug will "climb" the walls.

Writing Formulas

In "Which Formula?" you were shown six formulas and asked to select which ones could help you solve the problems given. Often in solving problems, these standard formulas won't do the job. When this happens, you have to write your own formula. You can do this by translating English sentences into mathematical sentences. Translate the following English sentence into a mathematical sentence. Use symbols and numerals for the words.

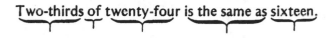

Two-thirds of twenty-four is the same as sixteen.

_____ __ _____ _____ _____

When you translate to a mathematical sentence to solve a problem, there will be an amount you don't know—the amount you are being asked to find. Use any letter (called a *variable*) for this amount and translate the following sentence:

Six times some number plus four is sixteen.

__ __ _____ __ __ __

A formula that has an "=" symbol is an equation. Solve the preceding equation; that is, find the value of the unknown number, the number represented by the letter you chose:_____

Now try this problem: Five more than some number equals one-half of twelve. What is your formula or equation?_____

What is the value of the unknown number?_____

Sometimes a problem can be rather complicated. When that happens, it's usually a good idea to write a simpler English sentence before you translate to a formula. Look at this problem:

The 150 cats at the Annual Dogpatch Cat Show were divided into two groups—long-haired cats and short-haired cats. There were twice as many short-haired as long-haired. How many of each kind of cat were there?

This looks pretty complicated, doesn't it? First, let's see what we know. If we add the number of long-haired cats and the number of short-haired cats, we'll get _____ . Also, _____ times the number of long-haired is the number of short-haired. So: Number of long-haired plus twice the number of long-haired is 150. Translate this to an equation:_____

Solve your equation:_____

How many of each kind of cat were there?_____

EXTENSION! Sometimes a problem contains more information than you need. Decide what you don't need in the following problem, write a formula, and solve the problem. Lisa's age is one-half that of her mother. Nine years ago, Lisa's age was only one-third of her mother's age. In 10 years, Lisa's mother will be 46 years old. How old is Lisa now?

Teacher's Notes for Writing Formulas

In this activity, the third in the formulas series, students are introduced to writing their own formulas. This, of course, means writing a mathematical equation to solve a word problem. By relating equation writing to the familiar work students have done in formulas, this activity makes translating English to mathematics easier to understand. Ask students to show their work.

NCTM Standards

1	2	3	4	5	6	7	8	9	10
•	•								•

Presenting the Activity

Emphasize the importance of thinking of mathematics as a language. Point out that you can't translate English into mathematics if you don't know the mathematical "words" for the English words. Mention, too, that a mathematical sentence can be read just as an English sentence is read. Students often think of equations and formulas as rows of symbols with no relationship to anything real. Emphasize that formulas are simply a shorthand way to write English—a way that makes relationships between quantities both more obvious and easier to work with.

Ask students for words or phrases that translate to $=$, $+$, $-$, $\times$, and $\div$ in mathematics. "Is," "is the same as," "equals," and "is equivalent to" are a few English phrases for the equals sign. Have students suggest four or five phrases for each of the other signs. They will encounter them in word problems and must know which sign is appropriate.

No student should have difficulty translating the first sentence to $2/3 \times 24 = 16$. Then discuss the idea of a variable. Students are familiar with letters in formulas, but assigning and inserting their own variables may seem strange. Tell them to think of them as strings tied around their fingers—a reminder that a number goes there.

The next sentence translates to $6x + 4 = 16$ and the value of x is 2. Remind students to check their answers to be sure six times two plus four is sixteen.

If some students have difficulty translating the next problem, have them write it as they wrote the preceding problem. (Most students should be able to write the equation without this step.) The equation is $5 + x = 1/2 \times 12$, so $x = 1$. Have students compare the English sentence to the mathematical formula. Ask them which one is easier to use to find the unknown number. It should be obvious to everyone that the formula is easier to work with, even though it says exactly the same thing as the English sentence. Students should now begin to see how mathematics can be useful in the physical world.

Often a complex physical situation can be reduced to a relatively simple mathematical equation. This equation can then be used to deduce information about the physical situation and to predict and interpret things about it.

The next problem will probably cause some initial difficulty for many students. Tell them that before they try to write an equation, they should do several things: They should read the problem through more than once, they should be sure they know what they are being asked to find, and they should study the problem to find out what information it gives them. *Thoroughly* discuss these ideas in relation to the problem given. The translated equation is $x + 2x = 150$ and x is 50. Of course, finding x doesn't finish the problem. Students must go back to the original problem to check what they were asked to find and what they did find. They found the number of long-haired cats, but still need to find the number of short-haired cats. They can find this number two ways: There are twice as many short-haired cats as long-haired cats, so $2 \times 50 = 100$ short-haired, or there are 150 cats at the show, so $150 - 50 = 100$ short-haired. This problem shows why it's important to write down what the variable stands for. Some students get to the end of a problem and don't know what they've found.

Students will need to work through more problems like the cat problem. Have them these problems:

1. Melvin bought two dozen apples at the store. One-fourth of the apples were green apples for a pie and the rest were red delicious apples. How many red apples did he buy? (18 red apples)
2. This year 359 dogs were entered in a dog show. This was four more than five times the number of dogs entered 10 years ago. How many dogs were entered 10 years ago? (71 dogs)

Extension

You don't need to know that Lisa's age was one-third that of her mother nine years ago. The equation for the problem is $2x = 46 - 10$ and x, Lisa's age, is 18. This is the easiest way to solve the problem. It could also be solved by excluding the information that Lisa's mother will be 46 years old in 10 years using the equation $2x = 3(x - 9) + 9$. Alternatively, you could exclude the information that Lisa's present age is one-half that of her mother and use the equation $3(x - 9) + 9 = 46 - 10$. Encourage your better students to find all three equations.

CHAPTER 5

Problem Solving in Other Areas

- What's It Really Cost?
- Successive Discounts
- Discounts and Increases ★
- Mathematics in Nature ★

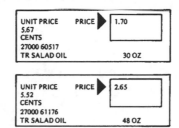

Most people who study mathematics study it with the primary intention of using mathematics in some pragmatic way in their careers. Less than 1% of your students will become mathematicians, but all of them will be consumers, and so this section should be of interest to most of your students and important to all of them. As will be apparent when you read the activities in this section, there is room for a lot more discussion of these topics than could possibly be covered in the individual activities' Teacher's Notes. You can take most of this group of activities as far as your students' interests and your time permit.

"What's It Really Cost?" deals with unit pricing. It is not a difficult skill to understand or acquire, but it does require some practice to attain speed and comfort. The activity gives various examples, first using exact computation and then with quick estimates. The Extension to this activity treats the topic much more broadly than the texts that include unit pricing. It explores the idea that unit price is not the sole criterion for a buy/don't buy decision.

"Successive Discounts" is especially pertinent in these days of rebates and discounts on discounts. The activities stress that whether you're considering automobiles, real estate, appliances, or insurance, it's usually a case of a professional seller, versus an amateur buyer. "Successive Discounts" thus treats percentages in a manner that is both realistic and motivating to students.

"Discounts and Increases" extends the concepts of "Successive Discounts" to allow the effects of simultaneous increases to be factored in. The activity provides solid practice in manipulating percentages. Perhaps more importantly, the Extension allows for a much deeper discussion of retail business operating realities than is encountered in most standard texts.

"Mathematics in Nature" will be especially appealing to your budding biology students. The Fibonacci sequence in nature often surprises students and its application is broad, not limited to just biology. The activity is a nice blend of pure and applied mathematics.

What's It Really Cost?

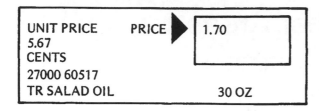

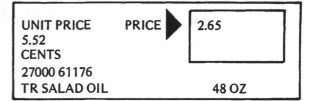

Look at the left-hand label above. How much does the bottle of salad oil cost? _____ What is its *unit price*? That is, how much does it cost *per ounce*? Look at the right-hand shelf label. What information is given on it?_____

_____

Which is the better buy? In other words, which has the lower unit price?_____ _____ Suppose the unit price were not given. How would you find it?

_____

When we talk about "price per ounce" or "price per pound," the "per" indicates a fraction. Fractions tell us that one number has to be divided by another number. If 30 oz of oil cost $1.70, then 1 oz costs $\frac{\$1.70}{30}$, or_____ ¢.

$1.83 $2.97

Compare two jars of the same brand of peanut butter: a 28-oz jar at $1.83 and a 48-oz jar at $2.97. Which is the better buy?_____
Rather than dividing, an easier method is to set up the two fractions as though they formed a proportion,

$$\frac{\$1.83}{28} > < \frac{\$2.97}{48}$$

$87.84 $83.16

Multiply as indicated by the arrows and compare the products. As you can see, the one on the left, the 28-oz jar, is more expensive.

This method is very easy if you have a calculator or when you are sitting with a pencil and paper, but it's not a very convenient method to use when you're in the aisle of a supermarket. Here you're much better off if you can round to numbers that give easy division problems. Then you can quickly get a close estimate of the unit prices.

Apple Juice	Price	Approx. Unit Price
Six 6-oz cans	$1.69	5¢
46-oz can	$1.25	3¢
32-oz bottle	$0.97	3¢

Tomato Juice	Price	Approx. Unit Price
Six 6-oz cans	$1.17	
12-oz can	$0.30	
24-oz can	$0.54	
46-oz can	$0.95	

Look at the prices in the left-hand table. For the apple juice, six 6-oz cans is about 35 oz. Five times 35 is 175. Because 169 is pretty close to 175, we see that buying the apple juice this way costs about 5¢ per ounce. Looking at the 46-oz can, we see that 3 times 45 is 135, so this comes to a little less than 3¢ per ounce. Three times 32 is 96, so the bottle is almost exactly 3¢ per ounce.

Now you make estimates of the cost per ounce for the different-sized cans of tomato juice.

EXTENSION! Smart shoppers know that although unit pricing is important, it is not the only factor you have to think about when buying. List the other things you'll want to think about before you buy.

Teacher's Notes for What's It Really Cost?

Although all of the activities in this series attempt to show mathematics as a subject that doesn't belong only between the covers of textbooks, this topic has immediate practical use for all your students. Simple multiplication and division and an acquaintance with rounding are the only prerequisite skills. Ask students to show their work.

───────────────────────── NCTM Standards ─────────────────────────

1	2	3	4	5	6	7	8	9	10
•	•		•				•	•	•

Presenting the Activity

The activity begins by asking students to look at some typical shelf labels. Although such labels have been used by large grocery stores for quite a while now, many people are still not aware of the information they contain nor are they in the habit of reading them.

The first shelf label shows that a 30-oz bottle of salad oil costs \$1.70. The unit price, or price per ounce, is 5.67¢. The second label shows that a 48-oz bottle of the same brand of salad oil costs \$2.65. The unit price of this bottle is 5.52¢. Thus, the 48-oz bottle has the lower unit price. (The numbers listed above the name of the product are standardized identification/inventory references.)

Most seventh or eighth grade texts devote a few pages to unit pricing and tell students to divide price by quantity. This is what the next question asks for.

By using this method, the students can find that the 48-oz jar of peanut butter is a slightly better buy at 6.19¢ per ounce compared to 6.54¢ per ounce for the 28-oz jar.

The next method—one that we haven't seen presented in standard texts—is undoubtedly the fastest method *when accuracy is required.* You may want to suggest or have your students suggest some additional examples to ascertain that the method works. Notice that the difference in the products reflects the difference in unit price. In this case, a quick inspection shows that 87.84 is about 5% greater than 83.16.

What is often needed, however, is a fast, easy method that a shopper can use on the spot (without the aid of a calculator). The degree of accuracy achieved by the earlier methods is usually not required. As students will discover in the Extension, there are several other criteria applied by intelligent shoppers, and unit price is often easily outweighed. For many classes, it is certainly worth your while to give students additional examples to practice estimating unit prices.

Although most students get the hang of unit prices without much trouble, estimating is a new activity to many of them. So it's a good idea to carefully walk them through the apple juice estimates before they are asked to estimate on their own. The tomato juice estimates should be $3\frac{1}{2}$¢, $2\frac{1}{2}$¢, 2¢, and 2¢.

Some students will ask themselves, "Why make such a big deal out of $1\frac{1}{2}$¢?" However $3\frac{1}{2}$¢ is 75% more than 2¢. The students can ask their parents what such a difference in food, clothing, housing, and transportation costs would mean to their budgets, lifestyles, and peace of mind.

Extension

Everyone in your class can enjoy, participate in, and profit from this discussion. The first topic you'll want to discuss is quality. Most standard textbooks treat unit pricing as though price per ounce were the only variable. However, quality variations do exist and shoppers must decide whether the additional quality is worth the additional price. (Your students should also be aware that many products are packaged under different labels with different prices. Even though the contents of the can, jar, or bottle are exactly the same, the price is different.)

While on the subject of quality, you may want to discuss buying quality that you don't need. It's foolish to buy picture-perfect tomatoes to put in spaghetti sauce or stew. New potatoes are just as good for mashing as the more expensive baking potatoes, and you certainly don't prepare your beef burgundy with an imported vintage wine.

A second topic is perishability or shelflife. Will the buyer use the entire package or end up throwing some of it away? A shopper can buy a 50 lb bag of potatoes at perhaps 20% less per pound than a 10 lb bag, but if 15 lb are thrown away, it isn't much of a bargain. If only one member of a household likes peanut butter or if all the members of the household use it only once in a while, they may decide that an added 5% isn't a bad price to pay for the added freshness they receive by buying smaller jars.

There's a third topic that some people don't like to admit as a valid criterion: Convenience. The 6-oz cans of apple juice cost 65% more per ounce than the larger sizes, but when packing lunches or going backpacking, the convenience of such packaging can be well worth the added cost. *The important point is for shoppers to know what they're paying for and then make the decision as to whether it's worth it.*

Finally, there are the cash-off coupons and manufacturers' refunds that can be clipped from newspapers and magazines. Used wisely, these can yield very significant savings. The trick, of course, is to use them only for items that you would buy anyway.

Successive Discounts

Last week Joanne saw a CD Walkman on sale at a department store for 30% off the list price. The list price is $32.00. Today she saw the same player on sale at a discount store. This store regularly sells everything at a 20% discount off the list price. The player was now on sale at 10% off the discount price. Which store, if either, do you think has the lower price for the player?_____
First, let's find the sale price at the department store. How can you do this? __

_____

At the discount store the regular price of the player is 20% off the list price. This means that 0.20 × list price = amount of discount. What is the amount of the discount?_____ What is the regular price at the discount store?___
_____ Finding the regular price this way took two steps-first finding the amount of the discount, and then subtracting this amount from the list price. We can find the regular price in one step if we use the equation

  0.80 × list price = regular price.

What single equation can you use now to find the sale price at the discount store?

_____

What is the sale price?_____ Which store had the lower price?_____
_____ It is also possible to find a single discount equivalent to the *successive discounts* of 20 and 10%. Examine the equations we just used:

   0.80 × list price = regular price,
  0.90 × regular price = sale price.

We can write these two equations as one equation:

  0.90 × (0.80 × list price) = sale price,
      0.72 × list price = sale price.

What is the single discount equivalent to the successive discounts of 20 and 10%?

_____

Here are the steps we used to change successive discounts to a single equivalent discount:

1. Change each of the successive discounts to decimal fractions. $20\% = 0.20; \ 10\% = 0.10$
2. Subtract each of these decimal fractions from 1. $1 - 0.20 = 0.80;$
   $1 - 0.10 = 0.90$
3. Multiply the results of step 2. $0.80 \times 0.90 = 0.72$
4. Subtract the result of step 3 from 1. $1 - 0.72 = 0.28$
5. Change the result of step 4 to percent form. $0.28 = 28\%$

What is the single discount equivalent to successive discounts of 30 and 5%?__

EXTENSION! A discount store regularly sells everything at a 10% discount off the list price. They have a clock radio on sale at 20% off their regular price. When it still doesn't sell, they discount it another 5% off the sale price. What is the single discount equivalent to the successive discounts of 10, 20, and 5%? The list price of the clock radio is $25.00. What is the final price?

Teacher's Notes for Successive Discounts

Many products, purchased primarily by teenagers, are often heavily advertised "sale items," so it is important for students to be knowledgeable shoppers. This activity, the following one, "Discounts and Increases," and "What's It Really Cost?" will help students acquire this knowledge and give them a better understanding of how retail businesses operate. In addition, problem-solving skills using percent and decimal fractions are developed.

Students should know how to multiply decimal fractions and should have had some practice in converting percents to decimals and decimals to percents. Ask students to show their work.

				NCTM Standards					
1	2	3	4	5	6	7	8	9	10
•	•		•				•	•	•

Presenting the Activity

Before beginning the student page, discuss retail business operations. Most students will know that stores make a profit by selling goods for more than they pay for them. Ask students for some of the reasons why stores have certain articles on sale (for example, because the product hasn't sold well at its regular price, because the article is no longer in style or popular, or to get customers into the store so they will buy other products). Ask why a discount store can regularly offer products at a lower price: they often buy in large quantities at a lower cost to the store; they usually have fewer employees than a department store; they spend less than a department store on furnishings, displays, and rent.

Now have students read through the problem at the top of the student page. They should answer the first question without doing any computation. Some students may think both stores offer the same discount. Point out that the discount store offers 10% off the already discounted price, *not* 10% off the list price.

The next series of questions leads students through the solution of the problem. Most students will find the sale price at the department store by first finding the amount of the discount,

$$30\% \times 32.00 = 0.30 \times 32.00 = 9.60,$$

and then subtracting this from the list price,

$$\$32.00 - \$9.60 = \$22.40.$$

A few students may realize that because the discount is 30%, the sale price is 70% of the list price, and find the sale price in one step. This method is discussed in the next paragraph.

The amount of the 20% discount at the discount store is $6.40, so the regular price is $25.60. This price can be found using one equation and students should be encouraged to use this method. To find the single equation for the sale price at the discount store, students must realize that a 10% discount off the regular price means the sale price is 90% of the regular price. That is,

90% × regular price = sale price

0.90 × $25.60 = $23.04.

Thus, the department store has the lower price. The single discount equivalent to the 20 and 10% successive discounts is 28%. Students can verify this by finding a 28% discount off $32.00.

You may want to have students consider a general method of converting any number of successive discounts to a single discount. Illustrate with two successive discounts of $d_1\%$ and $d_2\%$ operating on a price p. $p(1 - \frac{d_1}{100})$ represents the price after one discount has been computed; $p(1 - \frac{d_1}{100})(1 - \frac{d_2}{100})$ represents the price after the second discount has been computed; $(1 - \frac{d_1}{100})(1 - \frac{d_2}{100})$ represent the percentage that the new price is of the original price. Therefore, $1 - (1 - \frac{d_1}{100})(1 - \frac{d_2}{100})$ represents the discount taken off the original price to obtain the new price. Hence, successive discounts of $d_1\%$ and $d_2\%$ are equivalent to a single discount of $1 - (1 - \frac{d_1}{100})(1 - \frac{d_2}{100})$. If students have any difficulty understanding this algebraic development, have them substitute $d_1 = 20$, $d_2 = 10$, and $p = 32$ in the development. These are the values used in the problem they have already worked.

The series of steps on the student page is simply a verbal equivalent of the final algebraic expression. Using these steps, they will find that a 33.5% discount is equivalent to successive discounts of 30 and 5%. If necessary, give additional problems for more practice: What is the single discount equivalent to successive discounts of 20 and 15%? (32%) 40 and 10%? (46%) 20 and 50%? (60%).

Extension

Point out that the steps on the student page do not specify the number of successive discounts. Thus, they may be used for any number of successive discounts. You may want to have students verify this algebraically. The single discount equivalent to successive discounts of 10, 20, and 5% is found as follows:

1. 10% = 0.10; 20% = 0.20; 5% = 0.05

2. 1 − 0.10 = 0.90; 1 − 0.20 = 0.80; 1 − 0.05 = 0.95

3. (0.90)(0.80)(0.95) = 0.684

4. 1 − 0.684 = 0.316

5. 0.316 = 31.6%

The final price of the radio is 0.684 × $25 = $17.10.

Discounts and Increases

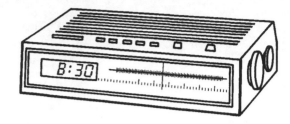

You've learned how to find a single discount equivalent to successive discounts and how to compare prices of products on sale. Now let's see how successive discounts and *increases* affect prices. Last year a digital clock radio sold at a list price of $30.00. Because of inflation, the list price of the radio has increased 15%. The radio is on sale at 20% off the present list price. Is the sale price greater or less than last year's list price?

Take a guess._____

First, let's find the new list price of the radio. How would you find the amount of the increase?_____

What is this amount?_____ What is the present list price of the radio?_____ What is the sale price?_____ How much more or less is this than last year's list price?_____

This problem shows a successive increase of 15% and discount of 20%. The effect of these two changes is a discount. We can find the single discount equivalent to this successive increase and discount.

1. Change each increase and discount
   to decimal fractions. $15\% = 0.15;\ 20\% = 0.20$
2. Subtract each discount from 1 and $1 + 0.15 = 1.15;$
   add each increase to 1. $1 - 0.20 = 0.80$
3. Multiply the results of step 2. $1.15 \times 0.80 = 0.92$
4. Find the difference between 1 and
   the result of step 3. $1 - 0.92 = 0.08$
5. Change the result of step 4 to percent form. $0.08\quad =\quad 8\%$

Notice that the result of step 3 is less than 1. This means the 8% is a discount. If the result of step 3 were greater than 1, we would subtract 1 from this result. The difference would then be an increase.

Try this problem: When the entrance price to a basketball game was decreased by 25%, the attendance at the game increased by 35%. What was the effect of these changes on the receipts?_____

EXTENSION! A store owner buys goods at a *cost price* that is lower than his selling price. The difference between the cost price and the selling price is the *markup* or *margin*. The markup is often written as a percent of the selling price. If you know the cost price and the desired margin, you can find the selling price:

$$\text{selling price} = \frac{\text{cost price}}{1 - \text{margin}}.$$

Suppose the cost price of a table is $18 and the store owner wants a margin of 40%. At what price should he sell the table? If the store owner puts the table on sale at a 10% discount, what will be the new margin?

Teacher's Notes for Discounts and Increases

This activity extends the ideas presented in "Successive Discounts" to include successive increases and discounts, and should follow it immediately. A technique for finding a single equivalent discount was developed in "Successive Discounts." This technique is now adapted to finding a single discount or increase equivalent to successive increases and discounts. Additional problem-solving skills using percent are also developed. Ask students to show their work.

				NCTM Standards					
1	2	3	4	5	6	7	8	9	10
•	•		•				•	•	•

Presenting the Activity

After students read through the problem at the top of the student page they should answer the first question without doing any computation. Most students will assume correctly that the sale price is less than last year's list price. The next series of questions verifies this and leads students to find the amount of the difference in price. To find the amount of the increase, students should find 15% of $30. This is $4.50, so the new list price is $34.50. The sale price is 80% of $34.50 or $27.60. Thus, the sale price is $2.40 less than last year's list price.

Now review the steps used to find a single discount equivalent to successive discounts as presented in "Successive Discounts."Have students compare these steps to the steps presented in this activity. Emphasize the changes in steps 2 and 4 that allow for increases. Students should verify that the 8% single discount is correct by finding that 8% of $30 is $2.40.

In solving the next problem, the student's work should be:

1. $25\% = 0.25$; $35\% = 0.35$

2. $1 - 0.25 = 0.75$; $1 + 0.35 = 1.35$

3. $(0.75)(1.35) = 1.0125$

4. $1.0125 - 1 = 0.0125$

5. $0.0125 = 1.25\%$

Thus, a successive discount of 25% and an increase of 35% is a single increase of 1.25%.

Discuss the ideas of supply and demand and other marketing factors. The preceding problem shows how the price of an item affects its sales—sometimes reducing the price will increase the sales enough to cause an overall increase in the amount of money brought in.

Present the following additional problems for more practice: When the price of a magazine was decreased by 15%, the sales increased by 20%. How were the receipts affected by these changes? (2% increase) What is the single discount or increase equivalent to a successive discount of 10% and increase of 10%? (1% discount) A successive increase of 10% and discount of 5%? (4.5% increase) Successive increases of 10 and 5% and discount of 15%? (1.825% discount) A successive increase of 10% and discounts of 10 and 5%? (5.95% discount).

Extension

The Extensions for many of the activities are not for everyone. This one, however, should be interesting to all students and should not be beyond their abilities. The discussion should bring out the point of view of the store owner as well as that of the consumer. Point out that the markup or margin is the owner's *gross profit*. Business expenses such as rent, salaries, taxes, utilities, and advertising are subtracted from the gross profit to give the *net profit*. Thus, a 40% margin is not necessarily a large one. Also, different types of businesses have different margins—luxury items such as jewelry have a higher markup than necessities such as food. (In fiscal 1980 only one of the country's six largest food store chains made more than 1¢ profit for each dollar of sales.) Ask students why they think this is so. The discussion should emphasize the importance of the relationship between prices and volume of sales.

The selling price of the table is found by using the equation on the student page:

$$\text{selling price} = \frac{\text{cost price}}{1 - \text{margin}}$$
$$= \frac{18}{1 - 0.40}$$
$$= \frac{18}{0.60}$$
$$= 30.$$

So, for a 40% margin, the selling price is $30.

If the table is on sale at a 10% discount, the sale price will be $0.90 \times \$30 = \$27$. To find the margin at this selling price, we again use the equation

$$\text{selling price} = \frac{\text{cost price}}{1 - \text{margin}}.$$

If we let $x = \text{margin}$,

$$27 = \frac{18}{1 - x}$$
$$27 - 27x = 18$$
$$27x = 9$$
$$x = 0.33\frac{1}{3} \quad \text{or} \quad 33\frac{1}{3}\%.$$

Thus, only a 10% discount on the price of the table reduces the margin from 40% to $33\frac{1}{3}\%$. Discuss how this will reduce the owner's net profit accordingly. Use this discussion to point out some of the difficulties in operating a retail business and why many small businesses fail each year: what will happen if too many items do not sell well and their prices must be reduced; how unexpected expenses such as a fire or burglary will decrease the net profit; the effects of shoplifting; the influence of location and advertising; and so on.

Mathematics in Nature

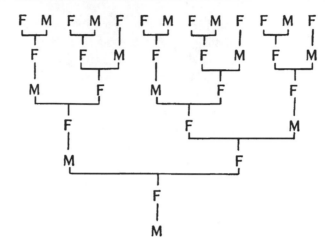

The family tree of a male bee shown above looks quite different from your family tree. You have two parents, they each have two parents (your grandparents), your grandparents each have two parents, and so on. A male bee has only one parent, a female bee. A female bee has two parents, a male and a female. Starting at the bottom of the family tree, count the number of bees in each generation. Write these numbers on the blanks below.

_____ , _____ , _____ , _____ , _____ , _____ , _____ , · · ·

This sequence of numbers is called the *Fibonacci sequence*. Each term in the sequence after the second term is the sum of the two previous terms. Write the next six terms of the sequence:

1, 1, 2, 3, 5, 8, 13, _____ , _____ , _____ , _____ , _____ , _____ , · · ·

Fibonacci numbers occur in nature in some unlikely situations—from pineapples to pinecones. One interesting occurrence of Fibonacci numbers is shown in the figure.

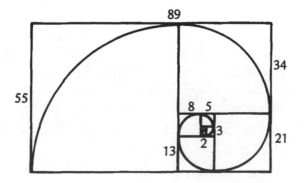

The large rectangle is 89 units long and 55 units wide. The rectangle is divided into a series of squares whose sides are Fibonacci numbers: 55, 34, 21, 13, 8, 5, 3, 2, 1, 1. The quarter circles in each square form a spiral curve. This curve is often found in the arrangements of seeds in flowers or in the shapes of seashells and snails.

The Fibonacci sequence has some interesting mathematical properties, too. Add the first six terms of the Fibonacci sequence. How does this sum compare to the eighth term?_____

Add the first seven terms and compare the sum to the ninth term:_____

Without adding, what is the sum of the first eleven terms?_____

Now look at this pattern of squares of Fibonacci numbers:

$$1^2 + 1^2 = 1 + 1 = 2 = 1 \times 2$$
$$1^2 + 1^2 + 2^2 = 1 + 1 + 4 = 6 = 2 \times 3$$
$$1^2 + 1^2 + 2^2 + 3^2 = 1 + 1 + 4 + 9 = 15 = 3 \times 5$$
$$1^2 + 1^2 + 2^2 + 3^2 + 5^2 = 1 + 1 + 4 + 9 + 25 = 40 = 5 \times 8$$

What is the next line of the pattern?_____

Without squaring or adding, what is the sum of the squares of the first eight Fibonacci numbers?_____

EXTENSION! Fibonacci numbers occur in plant growth. The numbers of petals of many flowers are Fibonacci numbers. They also occur in the arrangement of the leaves of a plant. Find some examples of these or study some pictures of them in botany books. Describe your results and tell why some plants grow in this pattern.

Teacher's Notes for Mathematics in Nature

Many students view mathematics as unrelated to both the real world and other fields of study. This activity should help dispel that view. As discussed in the activity, Fibonacci numbers occur in plant growth, sea creatures, and the reproduction of bees. This discussion of mathematics in the apparently unrelated field of biology should motivate students to look for mathematical concepts in other fields.

Additional examples of the Fibonacci sequence are given in both the algebra and geometry volumes of this series. Ask students to show their work.

				NCTM Standards					
1	2	3	4	5	6	7	8	9	10
•	•			•	•		•	•	•

Presenting the Activity

Students should have no difficulty understanding the family tree of the male bee. Be sure they start with the male bee at the bottom of the tree as one bee in the first generation. The number of bees in each generation is the Fibonacci sequence. The first 13 terms of the sequence are: 1, 1, 2, 3, 5, 8, 13, 21, 34, 55, 89, 144, 233. You may wish to have students extend the family tree one or two more generations to show that the pattern continues.

The next paragraph on the student page discusses some of the occurrences of Fibonacci numbers in nature. The scales of a pineapple spiral upward from the base in three distinct spirals: A group of 5 spirals winding gradually in one direction, a second group of 13 spirals winding more steeply in the same direction, and a third group of 8 spirals winding in the opposite direction. Each group of spirals consists of a Fibonacci number. Each pair of spirals interacts to give Fibonacci numbers.

The pinecone also presents a Fibonacci application. The bracts on the cone are considered to be modified leaves compressed into smaller space. There are two spirals, one to the left (clockwise) and the other to the right (counterclockwise). One spiral increases at a sharp angle, while the other increases more gradually. Counting the spirals produces Fibonacci numbers. For example, a white pinecone has five clockwise spirals and eight counter-clockwise spirals. Other pinecones have different Fibonacci ratios.

The rectangle on the student page is very close to the golden rectangle of the Greeks. This topic is discussed in detail in the geometry volume of this series. Its relationship to Fibonacci ratios is given in the Alternate Extension that follows.

The sum of the first six terms of the Fibonacci sequence is 20, one less than the eighth term. Similarly, the sum of the first seven terms is 33, one less than the ninth term. Students should see that the sum of the first 11 terms will be one less than the 13th term, or 232.

The next line of the pattern of squares is $1^2 + 1^2 + 2^2 + 3^2 + 5^2 + 8^2 = 1 + 1 + 4 + 9 + 25 + 64 = 104 = 8 \times 13$. Using this pattern, students can find the sum of the squares of the first eight Fibonacci numbers by multiplying the eighth term by the ninth term. The sum is 21×34 or 714.

Extension

The numbers of petals on flowers will vary according to the kind of flower and not every flower of a particular species is identical. The arrangement of leaves on a plant will usually produce two Fibonacci numbers: (1) the number of leaves it takes to go (rotating about the stem) from any given leaf to the next one "similarly placed" (above it and in the same direction) on the stem and (2) the number of revolutions as one follows the leaves in going from one leaf to another one "similarly placed." The following figure shows this leaf arrangement.

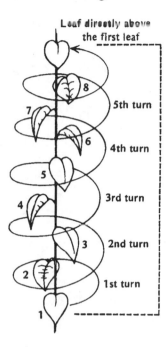

Alternate Extension

If researching plants would be difficult for many students, you may wish to present an alternate extension. Have students use graph paper to draw the rectangular figure on the student page. Then have them find the ratios to four decimal places of successive pairs of numbers in the Fibonacci sequence. These ratios and their decimal approximations are

$$\frac{1}{1} = 1.0000, \qquad \frac{2}{1} = 2.0000, \qquad \frac{3}{2} = 1.5000, \qquad \frac{5}{3} = 1.6667,$$

$$\frac{8}{5} = 1.6000, \qquad \frac{13}{8} = 1.6250, \qquad \frac{21}{13} = 1.6154, \qquad \frac{34}{21} = 1.6190.$$

$$\frac{55}{34} = 1.6176, \qquad \frac{89}{55} = 1.6182, \qquad \frac{144}{89} = 1.6180, \qquad \frac{233}{144} = 1.6181.$$

The ratios approach the ratio of the Greek golden rectangle: 1.61803 to 1.

CHAPTER 6

Recreational Mathematics

- Monday's Child ♦
- Palindromes ♦
- Odd-Order Magic Squares
- Even-Order Magic Squares
- Enrichment With a Hand-Held Calculator
- Scamps ★

1							
			3	62		14	
	2		32			17	
	29	4		20	61	36	
5			56				
		8	41		57	12	
					10		22

1	48	31	50	33	16	63	18
30	51	46	3	62	19	14	35
47	2	49	32	15	34	17	64
52	29	4	45	20	61	36	13
5	44	25	56	9	40	21	60
28	53	8	41	24	57	12	37
43	6	55	26	39	10	59	22
54	27	42	7	58	23	38	11

As Martin Gardner put it so well, "A good mathematical puzzle, paradox, or magic trick can stimulate a child's imagination much faster than a practical application (especially if the application is remote from the child's experience), and if the 'game' is chosen carefully it can lead almost effortlessly into significant mathematical ideas." For the reasons Gardner cites, these units should not be neglected or considered trivial, "merely recreational." Moreover, they should not be presented as a block, but rather interspersed throughout the year. They're especially valuable to use when interest in general seems to be waning.

"Monday's Child" is an activity that can be enjoyed by *all* of your students—even those who seem to be weak in mathematics. The motivation is built in and no mathematical skills are required. Moreover, it's just plainly a lot of fun. This is not to say that it's a dead end as far as your brighter students are concerned; the activity is an ideal springboard for studying why and how perpetual calendars can be developed.

"Palindromes," like the preceding activity, can appeal to a wide variety of students. The first half of the activity is within easy reach of all your students. The next 25% will cause difficulty only for a few students, and the Extension can give a good challenge to your better students. Throughout the activity, all of your students will be intrigued by the regular, really quite mechanical, process by which palindromes can be generated.

Magic squares have fascinated mathematicians and puzzle buffs for centuries. In the nineteenth century, one mathematician wrote a three-volume treatise on the subject. The two activities on magic squares that this book presents do more than just amuse. They offer a good opportunity to present sums of series, a tricky analysis of some number relationships, and, for chess fans, the knight's tour.

There is no question about the increasing importance of calculators and computers in not only mathematics education, but in everyday life for American citizens. "Enrichment With a Hand-Held Calculator" shows not only how calculators can relieve the tedium of many problems, it introduces the idea of averages and the arithmetic mean.

"Scamps" is very much in the vein of "Symmetric Multiplication," except that it has more surprises, is more fun. However, it is also somewhat more difficult than the others and you may wish to assign it only to your better students. You can be the best judge of its most appropriate use.

One Final Note

As the title of this series and several preceding statements indicate, these activities are meant to be motivational—to kindle interest—largely through the variety and novelty of problems that your students have not seen before. An interesting facet of motivation is how very contagious it is. Therefore, as a first cut, it is strongly recommended that you simply browse through this book and make a mental note of the activities that are especially appealing to you yourself. Then make the decision as to where in your course these activities fit best. Your own enthusiasm will assure a successful and enjoyable learning experience for your class.

Monday's Child

Monday's child is fair of face,
Tuesday's child is full of grace,
Wednesday's child is full of woe,
Thursday's child has far to go,
Friday's child is loving and giving,
Saturday's child has to work for its living,
But a child that's born on the Sabbath day
Is fair and wise and good and gay.

Do you know on what day of the week you were born? Many people don't. One way to find out is to find a calendar for the year you were born. However, there's an easier way—a perpetual calendar. This calendar is really a series of tables. You can use the tables to determine the day of the week for any date from 1600 A.D. to 2799 A.D.

July 4, 1776 was an important day in America's history. Using the accompanying tables, let's find out what day of the week it was.

1. Use Table 1 to find the first two digits of the year.
2. Use the bottom part of Table 2 to find the last two digits of the year.
3. Now use the top part of Table 2: Find the intersection of the row you found in step 1 with the column you found in step 2. This number is the code number for the year. What is the code number for the year 1776?_____
4. Find the code number for the year in the row of numbers at the top of Table 3 and the month in the column at the left.
5. Find the intersection of the row in which the month occurs and the column containing the code number for the year. This is the code number for the month. What is the code number for the month of July in 1776?_____
6. Find the code number for the month at the left of the top part of Table 4 and the numerical part of the date in the bottom part of Table 4.
7. Find the intersection of the row and column in step 6. This is the day of the week. What day of the week was July 4, 1776?_____

On what day of the week was Christmas in 1990?_____
What day of the week were you born?_____
Are you superstitious? You can use the perpetual calendar to find out when Fridays will be on the 13th of a month. Let's find out which months in 1989 had a Friday the 13th.

116

1. Find 13 at the bottom of Table 4.
2. Find Friday at the top of Table 4 in the column above the 13.
3. Find the code number at the left for the Friday in step 2. What is it?_____
4. Use Tables 1 and 2 to find the code number for the year. What is it?_____
5. Find the code number for the year at the top of Table 3.
6. Find the code number for Friday the 13th from step 3 in the column under the year's code number in Table 3.
7. Find the months listed at the left of Table 3. Which months in 1989 had a Friday the 13th?_____

Will there ever be a year without a Friday the 13th?_____

EXTENSION! You can also use the perpetual calendar to find the dates of all the Sundays in a particular month and year. Suppose the code number for the month of a particular year is 3. Find 3 to the left of Table 4 and go across the row to Sunday. The numbers in the column below Sunday at the bottom of Table 4 are 5, 12, 19, and 26. These are the dates of the four Sundays in the month and year you started with. On the third Sunday in July 1969, Neil Armstrong became the first man to walk on the moon. What was the date?

Teacher's Notes for Monday's Child

This activity is interesting to nearly all students. Although the topic is essentially recreational, it gives excellent practice in reading and understanding directions—an essential problem-solving skill. There are no mathematical prerequisites.

The perpetual calendar is shown on the page following these notes. A copy of it, as well as the student pages, must be provided to each student. Ask students to show their work.

					NCTM Standards					
1	2	3	4	5	6	7	8	9	10
•	•		•	•	•		•	•	•

Presenting the Activity

Begin with a brief history of our calendar and discuss the relationship of the calendar to astronomy. A year is defined as the interval of time between two passages of the earth through the same point in its orbit in relation to the sun. This is the solar year, 365.242216 mean solar days. The length of the year is not commensurable with the length of the day. Thus, much of the history of the calendar is the history of attempts to adjust these incommensurable units to obtain a simple and practical system.

The calendar story goes back to Romulus, the legendary founder of Rome, reputed to have introduced a year of 300 days divided into 10 months. His successor, Numa, added two months. Julius Caesar introduced a calendar based on a year containing 365.25 days. It provided a 366th day once every four years. The addition of leap years would completely compensate for the discrepancy if indeed the solar year consisted of exactly 365.25 days.

The difficulty with this method of reckoning is that 365.25 is not 365.242216. Although it may seem an insignificant quantity, the discrepancy accumulates over hundreds of years. By 1582 the accumulated error amounted to 10 days.

Pope Gregory XII tried to compensate for the error. Because the vernal equinox occurred on March 11 in 1582, he ordered that 10 days be dropped from the calendar dates in that year so that the vernal equinox would fall on March 21 as it should. When he proclaimed the calendar reform, he formulated the rules regarding the leap years. The Gregorian calendar has years (based on approximately 365.2425 days) divisible by four as leap years, unless they are divisible by 100 and not 400. Thus, 1700, 1800, 1900, and 2100 are not leap years, but 2000 is.

The change from the Julian to the Gregorian calendar was not made in Great Britain and its colonies until 1752. In September of that year, 11 days were dropped. The day after September 2 was September 14.

Now examine the perpetual calendar and the student pages. Although some students may have heard the verse at the top of the first page (anon.), most of them will be unable to apply it to themselves; that is, they will not know on what day of the week they were born. The verse should provide the motivation for studying the perpetual calendar.

Work through each step of the first problem with the students. It will be helpful to use an overhead projector to show the perpetual calendar as students work with their individual copies. The 17 of 1776 is in the second row of Table 1. The 76 is in the fifth column of Table 2. The intersection of the second row and the fifth column is the number 1. Thus, 1 is the code number for the year 1776. In Table 3, 1 is at the top of the first column and July is in the third row. Their intersection is 1, the code number for the month of July, 1776. In the top of Table 4, 1 is in the second row. In the numerical array at the bottom of Table 4, 4 is in the fourth column. The intersection of this row and column is Thursday. Thus, July 4, 1776 was on a Thursday.

Have students try to find what day of the week Christmas was in 1990. If necessary, work through the problem with them: The code number for 1990 is 0. The code number for December 1990 is 6. December 25, 1990 will be on Tuesday. Ask students to find other days, such as Christmas, Memorial Day, or Valentine's Day, of the present year. Then explain how to use the calendar to find Valentine's Day of a leap year, say 1988: The code number for 1988 is 5. The code number for February 1988 from Table 3 is 2. Call attention to the note on the perpetual calendar. Because 1988 is a leap year and we need the code number for February, we subtract 1 from the code number found in Table 3. Thus, the code number for February 1988 is 1 and Valentine's Day in 1988 was on Sunday.

Now, have students find what day of the week they were born.

The code number for any Friday the 13th is 0. The code number for 1989 is 6. Using Table 3, the 0 in the column under 6 is in the first row. Thus, in 1989 there was a Friday the 13th in January and October. You may want to give students additional practice in finding Friday the 13ths in other years. (Note: If the year is a leap year and the calendar shows Friday the 13ths in January or February, these months will *not* have them. However, there *will* be Friday the 13ths in the other months that are listed with January or February.)

There will never be a year without a Friday the 13th as is shown by 0 (the code number for Friday the 13th) being listed for at least one month under each of the headings of Table 3.

Extension

The code number for 1969 is 2. The code number for July 1969 is 2. The dates for the Sundays in July 1969 are the 6th, 13th, 20th, and 27th. Thus, the third Sunday in July 1969 was July 20, 1969.

This process can be used to find Thanksgiving (the fourth Thursday in November) or Labor Day (the first Monday in September) of any year.

Table 1

16	20	24
17	21	25
18	22	26
19	23	27

Table 2

6	0	1	2	3	4	5
4	5	6	0	1	2	3
2	3	4	5	6	0	1
0	1	2	3	4	5	6

00	01	02	03	—	04	05
06	07	—	08	09	10	11
—	12	13	14	15	—	16
17	18	19	—	20	21	22
23	—	24	25	26	27	—
28	29	30	31	—	32	33
34	35	—	36	37	38	39
—	40	41	42	43	—	44
45	46	47	—	48	49	50
51	—	52	53	54	55	—
56	57	58	59	—	60	61
62	63	—	64	65	66	67
—	68	69	70	71	—	72
73	74	75	—	76	77	78
79	—	80	81	82	83	—
84	85	86	87	—	88	89
90	91	—	92	93	94	95
—	96	97	98	99	—	—

Note: If the year in question is a leap year, subtract 1 from the code number found for the months of January and February in Table 3. If the month is not January or February, the code number found in Table 3 is correct.

Table 3

	1	2	3	4	5	6	0
Jan Oct	2	3	4	5	6	0	1
Feb Mar Nov	5	6	0	1	2	3	4
Apr July	1	2	3	4	5	6	0
May	3	4	5	6	0	1	2
June	6	0	1	2	3	4	5
Aug	4	5	6	0	1	2	3
Sept Dec	0	1	2	3	4	5	6

Table 4

0	Su	Mo	Tu	We	Th	Fr	Sa
1	Mo	Tu	We	Th	Fr	Sa	Su
2	Tu	We	Th	Fr	Sa	Su	Mo
3	We	Th	Fr	Sa	Su	Mo	Tu
4	Th	Fr	Sa	Su	Mo	Tu	We
5	Fr	Sa	Su	Mo	Tu	We	Th
6	Sa	Su	Mo	Tu	We	Th	Fr
	1	2	3	4	5	6	7
	8	9	10	11	12	13	14
	15	16	17	18	19	20	21
	22	23	24	25	26	27	28
	29	30	31	—	—	—	—

Palindromes

Think about the words *radar*, *civic*, and *rotator*. There's something unusual about them, isn't there? They're spelled backwards the same as they're spelled frontwards. These are called *palindromes*. There are even palindromic sentences, such as

Madam, I'm Adam.

Able was I ere I saw Elba.

Numbers, too, can be palindromes. Some examples are 747 and 95959. In this activity you can learn some of the properties of palindromes and how to form them. To make sure that you understand what palindromic numbers are, look at the following list of numbers and put an X next to the numbers that are palindromes.

	2312
	24542
	111011
	7171717
	63936
	84184

You can form palindromes by reversing the digits of a number and adding the number to its reversal. If this doesn't produce a palindrome, continue the process as follows. Let's take as a starting number 653:

$$653 + 356 = 1009,$$
$$1009 + 9001 = 10010$$
$$10010 + 01001 = 11011, \text{ a palindrome.}$$

Now you try it. Show your work.

<table>
<tr><td>753</td><td>421</td><td>877</td></tr>
</table>

Now form palindromes from the following sets of numbers.

554   752   653 | 780   483   681

What happened here? Can you state a rule to describe these situations?_____

_____

_____

_____

_____

EXTENSION! Form palindromes from 15 or 20 two-digit numbers. Can you then make a rule that tells you how many steps are required to form palindromes from two-digit numbers?

Teacher's Notes for Palindromes

Although this activity provides some good opportunities for multiplication practice, it's mostly just good fun. Thus, the activity makes a good motivator for students who aren't at the head of the class. At the same time, the Extension offers a rewarding challenge to anyone willing to invest an added 20 minutes or so. Ask students to show their work.

———————————————— NCTM Standards ————————————————

1	2	3	4	5	6	7	8	9	10
•					•				•

Presenting the Activity

A very few minutes is all that is needed to explain what a palindromic number is. In the examples given, the second, fourth, and fifth are palindromes, the first, third, and sixth are not. You may wish to ask your students how they could change these numbers to make them palindromes.

Discussing some of the properties of palindromes is optional. Palindromes may be either prime or composite. If they're prime, however, they contain an odd number of digits, with the exception of 11. There are an infinite number of palindromes that yield palindromic squares ($22^2 = 484$, $212^2 = 44944$, etc.). However, many palindromes do not have palindromic square roots. In general, numbers that yield palindromic cubes (some of which are prime and some of which are composite) are themselves palindromes. Also in general, I confess complete ignorance of the possible value of such information (either aesthetic or practical), but I do acknowledge that it's the sort of thing math educators like to include in their papers.

The heart of the activity is in generating palindromes by reversal and addition. The first three examples become palindromes via the following steps:

$$753 + 357 = 1110,$$
$$1110 + 0111 = 1221, \text{ a palindrome in two steps;}$$

$$421 + 124 = 545, \text{ a palindrome in one step;}$$

$$877 + 778 = 1655,$$
$$1655 + 5561 = 7216,$$
$$7216 + 6127 = 13343,$$
$$13343 + 34331 = 47674, \text{ a palindrome in four steps.}$$

The next two sets of numbers will present your students with something of a surprise.

If students aren't yet able to detect a pattern here, you may wish to provide a few additional examples. Then students should see that the first and last digits add up to the same number—9 in the first set and 7 in the second set.

```
   554        752        653
 + 455      + 257      + 356
  1009       1009       1009
+ 9001     + 9001     + 9001
 10010      10010      10010
+ 01001    + 01001    + 01001
 11011      11011      11011


   780        483        681
 + 087      + 384      + 186
   867        867        867
 + 768      + 768      + 768
  1635       1635       1635
+ 5361     + 5361     + 5361
  6996       6996       6996
```

If the middle digit is replaced, a different palindrome results, but for a given middle digit, the same palindrome will be achieved—and with the same steps—as long as the first and third digits add to the same number. Thus, if we substitute 6 as the middle digit in the first set, we find that 168, 267, 366, 465, 564, 663, 762, and 861 all yield the palindrome 13431 after three steps.

Extension

Students who understand the last problem will quickly look for a pattern that is related to the sums of the digits. They will discover that two-digit numbers whose sums are 11, 10, 12, 13, 14, 15, 16, or 17 yield palindromes after 1, 2, 2, 2, 3, 4, 6, and 24 steps, respectively. (*Note*: Numbers whose sums are less than 10 are palindromes after one step. Also, numbers such as 22, 33, and so on are already palindromes and are not included.) The tough one, of course, is the sum 17. Only two numbers have this digit sum: 89 and 98. These numbers can be either very challenging or very frustrating, depending on the tenacity of the student.

Odd-Order Magic Squares

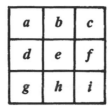 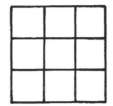

Have a look at the matrix on the left-hand side above. Could you substitute the numbers 1–9 for the letters *a–i* so that the sum of each row is the same? Moreover, make it so the sums of the columns and the diagonals all equal that same number. If you can do this, you will have constructed a *magic square*.

Do you think the job would be a little easier if we could find the sum of all the elements in a magic square (in this case, $1 + 2 + 3 + \ldots + 9$)? How would knowing this sum be of help?

Let's see how a young boy named Carl Friedrich Gauss figured this out more than 200 years ago. When Gauss was 10 years old he had a teacher who liked to give his students long, boring assignments. One day this teacher told the class to add the series $1 + 2 + 3 + \ldots + 99 + 100$. As the teacher finished stating the problem, young Gauss gave the answer. The amazed teacher asked him how he had found the answer so quickly. Gauss explained that rather than merely adding the 100 numbers in the order presented, he considered the following pairs: $1 + 100 = 101$; $2 + 99 = 101; 3 + 98 = 101; \ldots; 50 + 51 = 101$. Because there were 50 pairs of numbers, each with a sum of 101, his answer was $50 \times 101 = 5050$.

In our series of numbers (from 1 to 9), the sum of the first and last terms is_____ Now multiply this by the number of pairs of terms, $4\frac{1}{2}$. This gives us the sum of the elements in our magic square:_____
The three rows add up to the sum of the whole matrix. So the elements of each row must add up to_____. Similarly, the three columns add up to the sum of the whole matrix. Because the columns are equal, each column must add up to _____. This is why we asked if the job would be easier if we knew the sum of the whole matrix.

What number occupies the middle of a 3 × 3 magic square? That is, what number replaces *e*?_____ What makes you think so?_____.
The numbers that are added to 5 to form the middle row, middle column, and the diagonals must add up to 10. These numbers (1 and 9, 2 and 8, etc.) are called

complements. Note that the sum of the complements is the sum of the first and last numbers of the series. You now know enough to complete the magic square.

Can the number 1 occupy a corner of the magic square? Why or why not?____

_____

Can the number 3 be in the same row or column as 9? Why or why not?_____

_____

Once you've located 1, what other number is automatically located?_____

How about 3?_____ Now complete your magic square in the blank box at the beginning.

EXTENSION! Try to construct a 5×5 magic square using the principles you've learned with the 3×3 magic square.

Teacher's Notes for Odd-Order Magic Squares

To most young people, there's something very appealing and fascinating about magic squares. Very few books on recreational mathematics ignore them. Thus, magic squares provide an ideal vehicle for practice in good, head-scratching, problem solving. Your students should have been introduced to simple equations and should not be uncomfortable with seeing numerical values represented as variables. Ask students to show their work.

					NCTM Standards					
1	2	3	4	5	6	7	8	9	10	
•	•	•								

Presenting the Activity

It's a good idea to spend a minute or two to make sure that everyone understands what constitutes a row, a column, and a diagonal. Naturally, the students should realize that they're not to replace a–i with 1–9 in that order.

Point out that $a + b + c$ equals some number x. We know that $d + e + f$ equals x—otherwise it isn't a magic square. The same is true for $g + h + i$. Adding these, we get $3x$. This accounts for all the elements of the magic square, so if we know the sum of the numbers of the magic square, we know each row, column, or diagonal must be one-third that sum.

The Gauss anecdote is an interesting one. You may wish to have your students practice with other arithmetic series. The series needn't begin with 1, nor do they have to be consecutive integers. If your students are impressed because Gauss quickly figured this out at age 10, they should be.

Gauss' solution can be quickly seen if you visually show the linkages between terms in the series as shown in the figure

$$1 + 2 + 3 + 4 + 5 + 6 + 7 + 8 + 9$$

Then you can give the general statement: For the series $(1 + 2 + \cdots + n)$, the sum is $\frac{n}{2}(1 + n)$.

Many students will guess that 5 should occupy the middle square because it's the middle number in the series. Proving it, however, probably takes more ingenuity or mathematical sophistication than your students yet have. So don't let them dwell on the question too long before showing them a rather clever way to approach it. Show them the diagram

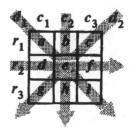

If we consider row 2, column 2, and the two diagonals, we've accounted for all of the outer elements once. However, we've included e four times. Looked at another way, we have the complete magic square plus three extra es. Each row, column, or diagonal equals 15, so $r_2 + c_2 + d_1 + d_2 = 60$. The magic square equals 45, so $3e = 15$ and $e = 5$. Have students enter 5 in the middle of the blank square.

Now consider the idea of complements. Because the sum of a diagonal is 15 and e equals 5, $a + i$ must equal 10. The same is true for $(c + g)$, $(b + h)$, and $(d + f)$.

Now lead your students through the following argument: The number 1 cannot occupy a corner position. Suppose $a = 1$; then $i = 9$. However, 2, 3, and 4 cannot be in the same row (or column) as 1, because there is no natural number less than 10 that would be large enough to occupy the third position of such a row (or column). This would leave only two positions to accommodate these three numbers (2, 3, and 4). Because this cannot be the case, the numbers 1 and 9 may occupy only the middle positions of a row (or column).

The number 3 cannot be in the same row (or column) as 9, because the third number in such a row (or column) would then have to be 3 to obtain the required sum of 15. This is not possible because a number can be used only once in the square. For the same reasons, neither 3 nor 7 may occupy corner positions.

There are eight variations, all of which are rotations or reflections of each other.

8	1	6
3	5	7
4	9	2

Extension

Begin by placing a 1 in the first position of the middle column. Continue by placing the next consecutive numbers successively in the cells of the (positive slope) diagonal. This, of course, is impossible because there are no cells "above" the square.

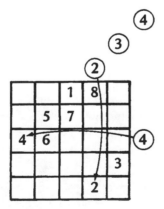

When a number must be placed in a position "above" the square, it should instead be placed in the last cell of the next column to the right. Then the next numbers are placed consecutively in this new (positive slope) diagonal. When (as in the accompanying figure) a number falls outside the square to the right, it should be placed in the first (to the left) cell of the next row above the row whose last (to the right) cell was just filled (as illustrated). The process then continues by filling consecutively on the new cell until an already occupied cell is reached (as is the case with 6 in the diagram). Rather than placing a second number in the occupied cell, the number is placed below the previous number. The process continues until the last number is reached.

After enough practice students will begin to recognize certain patterns (e.g., the last number always occupies the middle position of the bottom row). This is just one of many ways to construct odd-order magic squares. More adept students should be urged to justify this rather mechanical technique.

Even-Order Magic Squares

16	3	2	13
5	10	11	8
9	6	7	12
4	15	14	1

Albrecht Dürer, who lived in Germany from 1471 to 1528, is known primarily for his art. He also studied mathematics (as have many artists) and related much of it to his art. His engraving *Melencolia* was done in 1514. In the upper right corner of it there is a magic square. It may be one of the first magic squares to appear in Western civilization.

Notice that the date of the engraving is given by the bottom numbers of the two center columns. This square has many other interesting properties (in addition to the rows, columns, and diagonals having the same sum). List as many as you can.

_____

_____

Now let's try to construct some magic squares. Begin by simply writing 1–16 as shown on the left.

1	2	3	4
5	6	7	8
9	10	11	12
13	14	15	16

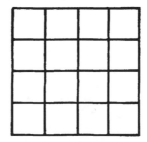

Exchange each number in the diagonals with its complement. (In this 4 × 4 magic square, the complements add up to_____.) So in the upper-left corner, instead of a 1, you'll place_____. Instead of a 4 in the upper-right corner, you'll place_____. Where the 6 now is, you'll place_____. Do the same with the rest of the diagonals.

Changing nothing else, check to see if you have constructed a magic square. What did Dürer do to obtain his magic square?_____

_____

To construct an 8 × 8 magic square, we can use the same procedure. First divide the 8 × 8 square into four 4 × 4 quadrants:

1	2	3	4	5	6	7	8
9	10	11	12	13	14	15	16
17	18	19	20	21	22	23	24
25	26	27	28	29	30	31	32
33	34	35	36	37	38	39	40
41	42	43	44	45	46	47	48
49	50	51	52	53	54	55	56
57	58	59	60	61	62	63	64

Using a separate sheet of paper, replace the numbers in the diagonals of the quadrants with their complements. The complements of an 8 × 8 magic square add up to _____. Does your new square work? Test it to be sure.

EXTENSION! There is a magic square sometimes called the *Knight's tour*. With this magic square you can trace the path of a knight moving on a chess board, reaching every square, but not landing on any square twice. The numbers show the order of the knight's moves yet still create a magic square. Your chess-playing friends can solve it with just the information given in the square

1							
			3	62		14	
	2		32			17	
	29	4		20	61	36	
5			56				
		8	41		57	12	
					10		22

But it's a much easier job if you apply the magic square properties you've just learned. (*Note*: A chess knight may move two squares across in either direction and one square up or down or it can move one square across and two squares up or down. The consecutive positions 1–5 in the diagram illustrate typical knight moves.)

Teachers Notes for Even-Order Magic Squares

This activity gives a lot of good practice in number operations and problem solving, but there are two other facets of this problem that motivate most students. First, the 4 × 4 magic square shown has an incredible number of interesting properties. Second, after having seen the rather painstaking procedure for constructing a 5 × 5 square, many students will be amazed by the simplicity of constructing an 8 × 8 square.

This activity should be studied after the students have finished "Odd-Order Magic Squares." Specifically, they should recall the idea of complements. Ask students to show their work.

―――――――――――――――――――― NCTM Standards ――――――――――――――――――――

1	2	3	4	5	6	7	8	9	10
•	•	•							

Presenting the Activity

You needn't be specific, but let your students know there are quite a few unusual properties to this magic square. Then allow the students several minutes of study so that they can give it their best shots. Then point out the following properties:

1. The four corner positions have a sum of 34.
2. The four corner 2 × 2 squares each have a sum of 34.
3. The center 2 × 2 square has a sum of 34.
4. The sum of the numbers in the diagonals equals the sum of those not in the diagonals.
5. The sum of the squares of the numbers in the diagonals (748) equals the sum of the squares of the numbers not in the diagonals.
6. The sum of the cubes of the numbers in the diagonals (9248) equals the sum of the cubes of the numbers not in the diagonals.
7. The sum of the squares of the numbers in the diagonals equals the sum of the squares of the numbers in the first and third rows (or columns), which equals the sum of the squares of the numbers in the second and fourth rows (or columns).
8. Note the following symmetries:

$$2 + 8 + 9 + 15 = 3 + 5 + 12 + 14 = 34;$$
$$2^2 + 8^2 + 9^2 + 15^2 = 3^2 + 5^2 + 12^2 + 14^2 = 374;$$
$$2^3 + 8^3 + 9^3 + 15^3 = 3^3 + 5^3 + 12^3 + 14^3 = 4624.$$

9. The sum of each upper and lower pair of numbers vertically or horizontally produces an interesting symmetry:

Vertically:	21	13	13	21
	13	21	21	13

Horizontally:	19	15
	15	19
	15	19
	19	15

When students replace the elements of the given square (consecutive 1–16) with their complements, they obtain the square

16	2	3	13
5	11	10	8
9	7	6	12
4	14	15	1

(Some students will notice that replacing with complements in this case is simply inverting the diagonals.) To obtain his square, Dürer switched columns 2 and 3. You may wish to point out that columns 1 and 4 can also be switched, and the square still "works." The same is true of rows 1 and 4 and of rows 2 and 3.

To construct an 8×8 magic square, divide the square into 4×4 magic squares and then replace the numbers in the diagonals of each of the 4×4 squares with their complements. Be certain your students understand they are to use the *diagonals of the quadrants*, not only the diagonals of the whole 8×8 square.

1	2	3	4	5	6	7	8
9	10	11	12	13	14	15	16
17	18	19	20	21	22	23	24
25	26	27	28	29	30	31	32
33	34	35	36	37	38	39	40
41	42	43	44	45	46	47	48
49	50	51	52	53	54	55	56
57	58	59	60	61	62	63	64

64	2	3	61	60	6	7	57
9	55	54	12	13	51	50	16
17	47	46	20	21	43	42	24
40	26	27	37	36	30	31	33
32	34	35	29	28	38	39	25
41	23	22	44	45	19	18	48
49	15	14	52	53	11	10	56
8	58	59	5	4	62	63	1

Extension

This is the sort of puzzle that can be fun in timed trials with math clubs—or chess clubs, in this case. Some students may ask to copy it to show friends or family.

1	48	31	50	33	16	63	18
30	51	46	3	62	19	14	35
47	2	49	32	15	34	17	64
52	29	4	45	20	61	36	13
5	44	25	56	9	40	21	60
28	53	8	41	24	57	12	37
43	6	55	26	39	10	59	22
54	27	42	7	58	23	38	11

Enrichment With a Hand-Held Calculator

Hand-held calculators have had a tremendous impact on the way arithmetic operations are performed in schools and offices. Calculators take out much of the boring drudgery that was once necessary for many problems. They can also be used just to play around with—as in riddles. "What must every man pay?" To find the answer, perform the calculation

$$2[60 - 0.243 + (12)(2400)] - 1.$$

The answer to the riddle is found by turning the calculator upside down.

"What did Eli Whitney's mother say the first time she saw him after he invented the cotton gin?" This time calculate

$$(9859.4875 \div 68.35) + 28.82734.$$

Again, turn the calculator upside down to read the answer.

Although calculators can perform arithmetic tasks for you, they can also teach you a lot about how our number system works. Take a page from a calendar. Select three consecutive days in a week and then pick the same three days for the following two weeks. This has already been done for you with the accompanying calendar page, but you can do the same thing with any calendar and any consecutive days and weeks.

S	M	T	W	T	F	S
1	2	3	4	5	6	7
8	9	10	11	12	13	14
15	16	17	18	19	20	21
22	23	24	25	26	27	28
29	30	31				

Take the smallest number (in this case 11), add 8 to it, and multiply by 9. What do you get?_____ Now add the numbers in either the middle row or column.

Multiply the sum by 3. What do you get?_____ Try it with some different months, days, and weeks. Do you get the same results?_____ Why?

_____

_____

_____

Let's take a look at another type of problem that we can explore with a calculator—though there are many, many more. Choose any three-digit number, say, 538, and punch it into your calculator. Now, *without hitting any operations buttons*, hit it again. Now your calculator will read 538538. Then divide by 7. Then divide the result by 11. Then by 13. What's your answer?_____ Try some other three-digit numbers. What are your answers and why does it work out this way?_____

_____

_____

EXTENSION! Select any three-digit number in which the hundreds digit and the units digit are unequal. Then write the number whose digits are in the reverse order from the selected number. Now subtract the smaller of these two numbers from the larger. Take the difference, reverse its digits, and add this "new" number to the original difference. What number do you always end up with? Why?

Teacher's Notes for Enrichment With a Hand-Held Calculator

When calculators are used in classrooms only for remedial drill work, the novelty soon wears off. Then the calculators are no longer an effective motivational device. A better use of calculators is as an aid to problem solving. Students can concentrate on interpreting the problem and not get bogged down in lengthy computation. Almost everyone makes an occasional error in computation involving large numbers and students become quickly discouraged if they spend 20 minutes solving a problem only to get a wrong answer because of a multiplication error. In this activity, the calculator provides ideas that serve as a springboard for additional activity. Thus, the calculator becomes a discovery and motivational device. Ask students to show their work.

				NCTM Standards					
1	2	3	4	5	6	7	8	9	10
•	•								

Presenting the Activity

Students should be familiar with the basic function of a calculator. For this activity, the calculators only need to have the four basic operations. It is not necessary to have a calculator for each student—students can work in pairs or small groups. The first two problems on the first student page are for practice and amusement. The answer to the first riddle is 57718.514. When students turn the calculator upside down, the answer reads "his bills." For the second riddle, the numerical answer is 173.07734; upside down, this reads "hello Eli." The following list shows the possible letters that can be made from upside down numerals:

0	1	2	3	4	5	6	7	8	9
O	I	Z	E	hr	S	g	L	B	b

Students can use this list to write their own riddles and problems. You may wish to wait until the activity is completed before giving students this list. Otherwise they may want to spend most of the class period making up riddles.

When students add 8 to 11 and multiply the sum by 9, they will get 171. They will also get 171 when they multiply the sum of the numbers in the middle row or column by 3. Have the students try this for other selections of nine numbers. They should realize that the sum of the numbers in the middle row or column multiplied by 3 is in fact the sum of the nine numbers. Students can verify this with their calculators.

You now have an excellent opportunity to investigate properties of the arithmetic mean. The middle number of the square of nine numbers is the arithmetic mean of the selected numbers. Thus, multiplying the middle number by 9 also gives the sum of the nine numbers (and adding 8 to the smallest number gives the middle number). Notice that although the computation here could be done without a calculator, using calculators allows students to focus all their attention on the discovery of the mathematical ideas.

Students will be quite surprised by the next problem. When they perform the operations on the second student page, they will get their original number back. Have them experiment with other numbers. If no one can figure out why it works, give them a hint: Ask what single operation can be used to replace the three divisions. They should realize that a single division by $7 \times 11 \times 13 = 1001$ was actually performed. Since $538 \times 1001 = 538538$, the puzzle is essentially solved. Students may wish to try this scheme on their calculators with other numbers.

Encourage students to try to establish other number patterns using their calculators to help with the computation.

Extension

Students will always end up with 1089. When they analyze the operations, they should notice that after subtracting, the ten's digit is always 9. Also, the sum of the hundred's and one's digits is always 9. Thus, their final sum must always be 1089. A mathematical justification for this follows.

The subtraction problem is written using a, b, and c as the digits,

$$
\begin{array}{r}
100a + 10b + c \\
- (100c + 10b + a)
\end{array}
$$

which is equal to

$$
\begin{array}{r}
100(a-1) + 10(b-1+10) \quad + (10+c) \\
- (100c \qquad\quad + 10b \qquad\qquad\quad + a \qquad\qquad) \\
\hline
100(a-1-c) + 10(b-1+10-b) + (10+c-a)
\end{array}
$$

Notice that because the one's digit is greater in the smaller number and the ten's digits are the same, it is necessary to "borrow" from both the ten's and hundred's digits of the larger number. If students have difficulty with this representation, have them work with an actual number to see what has been added to or subtracted from each digit. Now, the digits in the difference are reversed and added to the difference:

$$
\begin{array}{r}
100(a-1-c) + 10(9) + (10+c-a) \\
+ \quad 100(10+c-a) + 10(9) + (a-1-c) \\
\hline
1000 \ - \ 100 \ + \ 90 \ + \ 90 \ + \ 10 \ - \ 1 \ = \ 1089
\end{array}
$$

Thus, all the as, bs, and cs (the original digits) drop out, and we are left with 1089. Explain to students that this process is analogous to the preceding problem where the operations performed canceled each other. Essentially what we have here is the same sort of thing you have with the old puzzles such as, "Pick any number. Add 5. Multiply by 2. Subtract twice the original number. Divide by 10, etc. The answer is 1... ." The operations cancel out whatever original number was chosen.

137

Scamps

Those of you who have worked the Extension to "Enrichment With a Hand-Held Calculator" have run across the number 1089, a *scamp*. Scamps have been defined as "numbers that act kind of funny when they're multiplied or divided."

Let's see some other things that this scamp turns up when multiplied by perfectly innocent numbers. Perform the following multiplications and fill in the blanks:

$1089 \times 1 = $ _____ ,

$1089 \times 2 = $ _____ ,

$1089 \times 3 = $ _____ ,

$1089 \times 4 = $ _____ ,

$1089 \times 5 = $ _____ ,

$1089 \times 6 = $ _____ ,

$1089 \times 7 = $ _____ ,

$1089 \times 8 = $ _____ ,

$1089 \times 9 = $ _____ .

What kind of pattern can you find in your answers? _____

_____

_____

Why does this pattern occur? (To begin, ask yourself "What are the factors of 1089?") _____

_____

Now perform the following operations and try to explain the behavior of these scamps:

$1 \times 8 + 1 = $ _____ ,

$12 \times 8 + 2 = $ _____ ,

$123 \times 8 + 3 = $ _____ ,

$1234 \times 8 + 4 = $ _____ ,

$12{,}345 \times 8 + 5 = $ _____ ,

$123{,}456 \times 8 + 6 = $ _____ ,

$1{,}234{,}567 \times 8 + 7 = $ _____ ,

$12{,}345{,}678 \times 8 + 8 = $ _____ ,

$123{,}456{,}789 \times 8 + 9 = $ _____ ,

$$11 \times 11 = \underline{\hspace{5cm}},$$

$$111 \times 111 = \underline{\hspace{5cm}},$$

$$1111 \times 1111 = \underline{\hspace{5cm}},$$

$$11{,}111 \times 11{,}111 = \underline{\hspace{5cm}},$$

$$111{,}111 \times 111{,}111 = \underline{\hspace{5cm}},$$

$$1{,}111{,}111 \times 1{,}111{,}111 = \underline{\hspace{5cm}},$$

$$11{,}111{,}111 \times 11{,}111{,}111 = \underline{\hspace{5cm}},$$

$$111{,}111{,}111 \times 111{,}111{,}111 = \underline{\hspace{5cm}}.$$

Now use your calculator to find the decimal equivalents of the following (only the first six digits) fractions:

$$\frac{1}{7} = \underline{\hspace{2cm}}, \quad \frac{4}{7} = \underline{\hspace{2cm}}, \quad \frac{1}{13} = \underline{\hspace{2cm}}, \quad \frac{9}{13} = \underline{\hspace{2cm}},$$

$$\frac{2}{7} = \underline{\hspace{2cm}}, \quad \frac{5}{7} = \underline{\hspace{2cm}}, \quad \frac{3}{13} = \underline{\hspace{2cm}}, \quad \frac{10}{13} = \underline{\hspace{2cm}},$$

$$\frac{3}{7} = \underline{\hspace{2cm}}, \quad \frac{6}{7} = \underline{\hspace{2cm}}, \quad \frac{4}{13} = \underline{\hspace{2cm}}, \quad \frac{12}{13} = \underline{\hspace{2cm}}.$$

What sort of pattern do you find here? _____

_____

EXTENSION! Perform these operations and extend them to see if the pattern continues:

$$12321 = \frac{333 \times 333}{1+2+3+2+1} = \frac{\underline{\hspace{2cm}}}{9} = \underline{\hspace{3cm}},$$

$$1234321 = \frac{4444 \times 4444}{1+2+3+4+3+2+1} = \frac{\underline{\hspace{2cm}}}{16} = \underline{\hspace{3cm}}.$$

Teacher's Notes for Scamps

This activity presents an opportunity to explore number properties and examine how the decimal system works. Other activities of this type are "Multiplication—Still More Ways," "Palindromes," and "Symmetric Multiplication."

Although all the computation in this activity can be done without a calculator, slower students may find it some what tedious. Thus, if calculators are available, their use is recommended. Ask students to show their work.

					NCTM Standards				
1	2	3	4	5	6	7	8	9	10
•	•								

Presenting the Activity

The products of 1089 are

$$1089 \times 1 = 1089,$$
$$1089 \times 2 = 2178,$$
$$1089 \times 3 = 3267,$$
$$1089 \times 4 = 4356,$$
$$1089 \times 5 = 5445,$$
$$1089 \times 6 = 6534,$$
$$1089 \times 7 = 7623,$$
$$1089 \times 8 = 8712,$$
$$1089 \times 9 = 9801.$$

In the products, students should notice the symmetry of the first two columns and the last two columns. In the first two columns the digits increase by one in each new product. In the last two columns the digits decrease by one in each new product. Notice too, the reversal of the digits of the last product and the first product.

To discover why this pattern occurs, students should factor 1089:

$$1089 = 11 \times 11 \times 9$$
$$= 11(10 + 1)(10 - 1)$$
$$= 11(100 - 1)$$
$$= 1100 - 11,$$

$$1089 \times 2 = 2200 - 22 = 2178,$$
$$1089 \times 3 = 3300 - 33 = 3267, \text{ and so on.}$$

The answers for the next two patterns are

$$1 \times 8 + 1 = 9,$$
$$12 \times 8 + 2 = 98,$$
$$123 \times 8 + 3 = 987,$$
$$1234 \times 8 + 4 = 9876,$$
$$12{,}345 \times 8 + 5 = 98{,}765,$$
$$123{,}456 \times 8 + 6 = 987{,}654,$$
$$1{,}234{,}567 \times 8 + 7 = 9{,}876{,}543,$$
$$12{,}345{,}678 \times 8 + 8 = 98{,}765{,}432,$$
$$123{,}456{,}789 \times 8 + 9 = 987{,}654{,}321$$

$$11 \times 11 = 121,$$
$$111 \times 111 = 12{,}321,$$
$$1111 \times 1111 = 1{,}234{,}321,$$
$$11{,}111 \times 11{,}111 = 123{,}454{,}321,$$
$$111{,}111 \times 111{,}111 = 12{,}345{,}654{,}321,$$
$$1{,}111{,}111 \times 1{,}111{,}111 = 1{,}234{,}567{,}654{,}321,$$
$$11{,}111{,}111 \times 11{,}111{,}111 = 123{,}456{,}787{,}654{,}321,$$
$$111{,}111{,}111 \times 111{,}111{,}111 = 12{,}345{,}678{,}987{,}654{,}321.$$

For the first pattern, consider

$$
\begin{aligned}
123456 \times 8 + 6 &= 123456(10 - 2) + 6 \\
&= 1234560 - 2(123456) + 6 \\
&= 1234566 - 123456 - 123456 \\
&= 1111110 - 123456 \\
&= 987654.
\end{aligned}
$$

For the second pattern, write each factor as a sum. For example,

$$
\begin{aligned}
111 \times 111 &= (100 + 10 + 1)(100 + 10 + 1) \\
&= 10000 + 2000 + 300 + 20 + 1 \\
&= 12321.
\end{aligned}
$$

Each fraction produces a repeating decimal:

$$\frac{1}{7} = 0.\overline{142857}, \qquad \frac{1}{13} = 0.\overline{076923},$$
$$\frac{2}{7} = 0.\overline{285714}, \qquad \frac{3}{13} = 0.\overline{230769},$$
$$\frac{3}{7} = 0.\overline{428571}, \qquad \frac{4}{13} = 0.\overline{307692},$$
$$\frac{4}{7} = 0.\overline{571428}, \qquad \frac{9}{13} = 0.\overline{692307},$$

141

$$\frac{5}{7} = 0.\overline{714285}, \qquad \frac{10}{13} = 0.\overline{769230},$$

$$\frac{6}{7} = 0.\overline{857142}, \qquad \frac{12}{13} = 0.\overline{923076}.$$

Consider the digits in the decimal as "circular." That is, for the sevenths, the 4 always follows the 1, the 2 follows the 4, and so on. As the fractions increase, the first digit of the decimal increases. Once the first digit of the decimal is determined, the remaining digits follow in order.

Students may wish to explore the other fractions with denominator 13. They will find a similar pattern. Suggest denominators of 37, too.

Extension

To show why this pattern works, consider the equation

$$\frac{333 \times 333}{1+2+3+2+1} = \frac{3 \times 3 \times 111 \times 111}{3 \times 3} = 111 \times 111.$$

This is simply the product students found previously on the student page.

Resource A: List of Activities and the NCTM Standards Addressed by Each

Name of Activity	NCTM Standards									
	1	2	3	4	5	6	7	8	9	10
The Fascinating Number 9	•	•				•	•	•		
Symmetric Multiplication	•	•								
Multiplication—Still More Ways	•	•				•			•	
Divisibiliti	•					•			•	
Triangular Numbers	•	•	•			•				
Prime After Prime	•	•				•			•	
Geometric Dissections		•	•	•		•	•			
How Many Colors?		•	•			•		•	•	•
To Stretch a Point			•							
You Can't Get There From Here			•			•			•	
The Moebius Strip			•	•	•	•		•	•	•
The Tower of Hanoi		•	•	•	•	•				•
The Game of Nim	•	•			•	•				
A Checkerboard Calculator	•	•	•	•		•		•	•	
The Googol	•	•				•		•	•	•
Posers		•			•	•		•		•
The Euclidean Algorithm	•	•								
Formulas—Mathematical Shorthand		•		•	•				•	•
Which Formula?		•	•	•	•					•
Writing Formulas	•	•								•
What's It Really Cost?	•	•		•				•	•	•
Successive Discounts	•	•		•				•	•	•
Discounts and Increases	•	•		•				•	•	•
Mathematics in Nature	•	•			•	•		•	•	•
Monday's Child	•	•		•	•	•		•	•	•
Palindromes	•					•				•
Odd-Order Magic Squares	•	•	•							
Even-Order Magic Squares	•	•	•							
Enrichment With a Hand-Held Calculator	•	•								
Scamps	•	•								

CORWIN
PRESS

The Corwin Press logo—a raven striding across an open book—represents the happy union of courage and learning. We are a professional-level publisher of books and journals for K-12 educators, and we are committed to creating and providing resources that embody these qualities. Corwin's motto is "Success for All Learners."

Printed in the United States
By Bookmasters